Instructor's Resource Manual
Volume 2: Printed Test Bank

Kevin M. Bodden
Lewis and Clark Community College

Christopher J. Rigdon
Southern Illinois University, Edwardsville

Calculus
NINTH EDITION

Varberg   Purcell   Rigdon

Upper Saddle River, NJ 07458

Editor-in-Chief: Sally Yagan
Acquisitions Editor: Adam Jaworski
Project Manager: Dawn Murrin
Editorial Assistant: Christine Whitlock
Executive Managing Editor: Kathleen Schiaparelli
Senior Managing Editor: Nicole M. Jackson
Assistant Managing Editor: Karen Bosch Petrov
Production Editor: Jessica Barna
Supplement Cover Manager: Paul Gourhan
Supplement Cover Designer: Christopher Kossa
Manufacturing Buyer: Ilene Kahn
Manufacturing Manager: Alexis Heydt-Long

© 2007 Pearson Education, Inc.
Pearson Prentice Hall
Pearson Education, Inc.
Upper Saddle River, NJ 07458

All rights reserved. No part of this book may be reproduced in any form or by any means, without permission in writing from the publisher.

Pearson Prentice Hall™ is a trademark of Pearson Education, Inc.

The author and publisher of this book have used their best efforts in preparing this book. These efforts include the development, research, and testing of the theories and programs to determine their effectiveness. The author and publisher make no warranty of any kind, expressed or implied, with regard to these programs or the documentation contained in this book. The author and publisher shall not be liable in any event for incidental or consequential damages in connection with, or arising out of, the furnishing, performance, or use of these programs.

> **This work is protected by United States copyright laws and is provided solely for the use of instructors in teaching their courses and assessing student learning. Dissemination or sale of any part of this work (including on the World Wide Web) will destroy the integrity of the work and is not permitted. The work and materials from it should never be made available to students except by instructors using the accompanying text in their classes. All recipients of this work are expected to abide by these restrictions and to honor the intended pedagogical purposes and the needs of other instructors who rely on these materials.**

Printed in the United States of America

10 9 8 7 6 5 4 3 2 1

ISBN 0-13-146965-7

Pearson Education Ltd., *London*
Pearson Education Australia Pty. Ltd., *Sydney*
Pearson Education Singapore, Pte. Ltd.
Pearson Education North Asia Ltd., *Hong Kong*
Pearson Education Canada, Inc., *Toronto*
Pearson Educación de Mexico, S.A. de C.V.
Pearson Education—Japan, *Tokyo*
Pearson Education Malaysia, Pte. Ltd.

Calculus
NINTH EDITION

Varberg   Purcell   Rigdon

Printed Test Bank

0. Preliminaries   E1
1. Limits   E21
2. The Derivative   E33
3. Applications of the Derivative   E50
4. The Definite Integral   E74
5. Applications of the Integral   E90
6. Transcendental Functions   E102
7. Techniques of Integration   E122
8. Indeterminate Forms and Improper Integrals   E136
9. Infinite Series   E141
10. Conics and Polar Coordinates   E163
11. Geometry in Space and Vectors   E186
12. Derivatives For Functions of Two or More Variables   E206
13. Multiple Integrals   E218
14. Vector Calculus   E232
15. Differential Equations   E241

0. Preliminaries

1. The sum of a rational number and an irrational number must be _____.

   Answer: irrational

   Difficulty: 2   Section: 1

2. Give an example of irrational numbers whose product is irrational and an example whose product is rational.

   Answer: Varies. For the first question, $\sqrt{2}$ and $\sqrt{3}$ work, and for the second question $\sqrt{2}$ and $\sqrt{8}$ work.

   Difficulty: 2   Section: 1

3. Find a rational number between $\frac{3}{4}$ and $\frac{4}{5}$.

   Answer: Varies. The average is $\frac{31}{40}$.

   Difficulty: 1   Section 1

4. Change the repeating decimal $1.163636363\ldots$ to a ratio of two integers.

   Answer: $\frac{64}{55}$.

   Difficulty: 2   Section: 1

5. Find the best decimal approximation to $\dfrac{\sqrt{959} - 5^{2.3}}{5^4 - \sqrt{37.912}}$ that your calculator allows.

   Answer: Depends on calculator. About -0.0154299.

   Difficulty: 2   Section 1

6. Find the solution set for the inequality $2x - 11 < 4x - 1$.

   Answer: $(-5, \infty)$.

   Difficulty: 1   Section: 2

7. Find the solution set for the inequality $5x - 11 < 15x - 22$.

   Answer: $\left(\dfrac{11}{10}, \infty\right)$.

   Difficulty: 1   Section: 2

8. Find the solution set for the inequality $3x - 10 > 5x + 2$.

   Answer: $(-\infty, -6)$

   Difficulty: 1   Section: 2

9. Find the solution set for the inequality $x^2 - 9x + 20 > 0$.

   Answer: $(-\infty, 4) \cup (5, \infty)$.

   Difficulty: 2   Section: 2

10. Find the solution set for the inequality $(x-1)^2 \geq 16$.

    Answer: $(-\infty, -3] \cup [5, \infty)$.

    Difficulty: 2   Section: 2

11. Find the solution set for the inequality $x^2 + x - 2 < 0$.

    Answer: $(-2, 1)$.

    Difficulty: 2   Section: 2

12. Find the solution set for the inequality $x^2 - x - 8 < 4$.

    Answer: $(-3, 4)$.

    Difficulty: 2   Section: 2

13. Find the solution set for the inequality $(x-5)(x-2) \geq -2$.

    Answer: $(-\infty, 3] \cup [4, \infty)$.

    Difficulty: 2   Section: 2

14. Find the solution set for the inequality $\dfrac{x+1}{x-4} \leq 2$.

    Answer: $(-\infty, 4) \cup [9, \infty)$.

    Difficulty: 2   Section: 2

15. Find the solution set for the inequality $\dfrac{x+1}{x-6} < 0$.

    Answer: $(-1, 6)$.

    Difficulty: 2   Section: 2

16. Find the solution set for the inequality $\dfrac{x-4}{x+5} \leq 0$.

    Answer: $(-5, 4]$.

    Difficulty: 2   Section: 2

17. Find the solution set for the inequality $\dfrac{x+1}{4x-3} > 0$.

    Answer: $(-\infty, -1) \cup \left(\dfrac{3}{4}, \infty\right)$.

    Difficulty: 2   Section: 2

18. Find the solution set for the inequality $\dfrac{3x-2}{x-2} \leq 1$.

    Answer: $[0, 2)$.

    Difficulty: 2 Section: 2

19. Find the solution set for the inequality $\dfrac{(x-3)^2}{2x+1} > 0$.

    Answer: $\left(-\dfrac{1}{2}, \infty\right)$.

    Difficulty: 2 Section: 2

20. Find the solution set for the inequality $|3x+1| \leq 7$.

    Answer: $\left[-\dfrac{8}{3}, 2\right]$.

    Difficulty: 2 Section: 2

21. Find the solution set for the inequality $|10x+17| < 45$.

    Answer: $\left(-\dfrac{31}{5}, \dfrac{14}{5}\right)$.

    Difficulty: 2 Section: 2

22. Find the solution set for the inequality $|3x+16| < 8$.

    Answer: $\left(-8, -\dfrac{8}{3}\right)$.

    Difficulty: 2 Section: 2

23. Find the solution set for the inequality $|5x+17| < 7$.

    Answer: $\left(-\dfrac{24}{5}, -2\right)$.

    Difficulty: 2 Section: 2

24. Describe the interval $[-5, 5]$ by means of an inequality involving absolute values.

    Answer: $|x| \leq 5$.

    Difficulty: 1 Section: 2

25. Describe the interval $(0, 6)$ by means of an inequality involving absolute values.

    Answer: $|x-3| < 3$.

    Difficulty: 1 Section: 2

26. Describe the interval $(-2, 4)$ by means of an inequality involving absolute values.

Answer: $|x - 1| < 3$.

Difficulty: 1   Section: 2

27. Find the distance between the points $(-1, 1)$ and $(2, -1)$.

    Answer: $\sqrt{13}$.

    Difficulty: 1   Section: 3

28. Find the distance between the points $(-4, 6)$ and $(-3, -1)$.

    Answer: $5\sqrt{2}$.

    Difficulty: 1   Section: 3

29. Find the distance between the points $(-5, -1)$ and $(3, 1)$.

    Answer: $2\sqrt{17}$.

    Difficulty: 1   Section: 3

30. Find the distance between the points $(3, 6)$ and $(1, -5)$.

    Answer: $5\sqrt{5}$.

    Difficulty: 1   Section: 3

31. Find the distance between the points $(1, -1)$ and $(3, 1)$.

    Answer: $2\sqrt{2}$.

    Difficulty: 1   Section: 3

32. Write the equation of the circle with radius 7 and center $(-3, 7)$.

    Answer: $(x + 3)^2 + (y - 7)^2 = 49$.

    Difficulty: 1   Section: 3

33. Write the equation of the circle with radius 1 and center $(0, 1)$.

    Answer: $x^2 + (y - 1)^2 = 1$.

    Difficulty: 1   Section: 3

34. Write the equation of the circle with radius 2 and center $(2, -3)$.

    Answer: $(x - 2)^2 + (y + 3)^2 = 4$.

    Difficulty: 1   Section: 3

35. Write the equation of the circle with radius $\dfrac{3}{4}$ and center $(0, 0)$.

    Answer: $x^2 + y^2 = \dfrac{9}{16}$.

Difficulty: 1   Section: 3

36. Write the equation of the circle with radius 3 and center $(-1, 3)$.

    Answer: $(x+1)^2 + (y-3)^2 = 9$.

    Difficulty: 1   Section: 3

37. Write the equation of the circle with diameter AB for $A(-4, 2)$ and $B(2, -2)$.

    Answer: $(x+1)^2 + y^2 = 13$.

    Difficulty: 2   Section: 3

38. Give the center and radius of the circle $2x^2 + 2y^2 = 4y - 6x - 6$.

    Answer: $\left(-\frac{3}{2}, 1\right), \frac{1}{2}$.

    Difficulty: 2   Section: 3

39. Give the center and radius of the circle $3x^2 + 3y^2 - 12x = 6 - 8y$.

    Answer: $\left(2, -\frac{4}{3}\right), \frac{\sqrt{70}}{3}$.

    Difficulty: 2   Section: 3

40. Give the center and radius of the circle $3x^2 + 3y^2 - 12x + 78y + 492 = 0$.

    Answer: $(2, -13), 3$.

    Difficulty: 2   Section: 3

41. Show that the locus of $x^2 + y^2 - 2x + 2y + 2 = 0$ is a degenerate circle.

    Answer: The equation is $(x-1)^2 + (y+1)^2 = 0$.

    Difficulty: 2   Section: 3

42. Show that the locus of $x^2 + y^2 + 2x - 2y + 7 = 0$ is empty.

    Answer: The equation is $(x+1)^2 + (y-1)^2 = -5$.

    Difficulty: 2   Section: 3

43. Write the equation of the line through the points $(0, 1)$ and $(2, 3)$.

    Answer: $y = x + 1$.

    Difficulty: 1   Section: 3

44. Write the equation of the line through the points $(-3, 2)$ and $(0, 3)$.

    Answer: $y = \frac{1}{3}x + 3$.

Difficulty: 1   Section: 3

45. Write the equation of the line through the points $(2, -3)$ and $(1, -2)$.

    Answer: $y = -x - 1$.

    Difficulty: 1   Section: 3

46. Write the equation of the line through the points $(0, 3)$ and $(1, 0)$.

    Answer: $y = -3x + 3$.

    Difficulty: 1   Section: 3

47. What is the slope of the line with equation $2x - 3y = 5$?

    Answer: $m = \dfrac{2}{3}$.

    Difficulty: 1   Section: 3

48. What is the slope of the line with equation $\dfrac{x}{5} + \dfrac{y}{4} = 5$?

    Answer: $m = -\dfrac{4}{5}$.

    Difficulty: 1   Section: 3

49. What is the slope of the line with equation $-3x + 4y = 7$?

    Answer: $m = \dfrac{3}{4}$.

    Difficulty: 1   Section: 3

50. What is the slope of the line with equation $\dfrac{x}{2} + \dfrac{y}{3} = 2$?

    Answer: $m = -\dfrac{3}{2}$.

    Difficulty: 1   Section: 3

51. Find the equation of the vertical line passing through $(3, 2)$.

    Answer: $x = 3$.

    Difficulty: 1   Section: 3

52. Find the equation of the vertical line passing through $(-1, 0)$.

    Answer: $x = -1$.

    Difficulty: 1   Section: 3

53. Find the equation of the vertical line passing through $(-6, 7)$.

Answer: $x = -6$.

Difficulty: 1   Section: 3

54. Write an equation of the line that is parallel to $-3x - 2y + 2 = 0$ and that passes through $(4, 3)$.

    Answer: $y = -\dfrac{3}{2}x + 9$.

    Difficulty: 2   Section: 3

55. Write an equation of the line that is parallel to $2y - 3x = 0$ and that passes through $(-4, -7)$.

    Answer: $y = \dfrac{3}{2}x - 1$.

    Difficulty: 2   Section: 3

56. Write an equation of the line that is parallel to $-3x + y = 0$ and that passes through $\left(\dfrac{1}{3}, -2\right)$.

    Answer: $y = 3x - 3$.

    Difficulty: 2   Section: 3

57. Write an equation of the line that is parallel to $x - 3y = 3$ and that passes through $(6, -5)$.

    Answer: $y = \dfrac{1}{3}x - 7$.

    Difficulty: 2   Section: 3

58. Write an equation of the line perpendicular to $5x + 2y = -4$ and passing through $(9, -4)$.

    Answer: $y = \dfrac{2x}{5} - \dfrac{38}{5}$.

    Difficulty: 2   Section: 3

59. Write an equation of the line perpendicular to $\dfrac{x}{8} + \dfrac{y}{2} = 1$ and passing through $(-5, 4)$.

    Answer: $y = 4x + 24$.

    Difficulty: 2   Section: 3

60. Write an equation of the line perpendicular to $2x + 3y = 7$ and passing through $(4, 7)$.

    Answer: $y = \dfrac{3}{2}x + 1$.

    Difficulty: 2   Section: 3

61. Write an equation of the line perpendicular to $-3x = 7y + 2$ and passing through $(3, 1)$.

    Answer: $\dfrac{7}{3}x - 6$.

Difficulty: 2   Section: 3

62. Sketch the graph of $x = 2 + y^2$.

    Answer:

    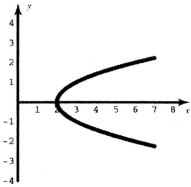

    Difficulty: 1   Section: 4

63. Sketch the graph of $4x^2 + y^2 = 4$.

    Answer:

    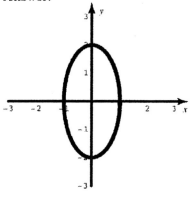

    Difficulty: 2   Section: 4

64. Sketch the graph of $y = \sqrt[3]{x}$.

    Answer:

    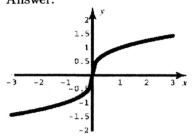

    Difficulty: 2   Section: 4

65. Sketch the graph of $y = \dfrac{1}{x}$.

Answer:

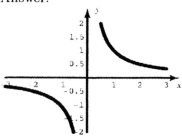

Difficulty: 2   Section: 4

66. Sketch the graph of $y = (x-1)(x+2)(x-3)$.

Answer:

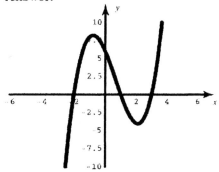

Difficulty: 3   Section: 4

67. Sketch the graphs of $y = x^2 + 2x + 1$ and $y = x + 3$ on the same coordinate plane and give the points of intersection.

Answer: Points are $(-2, 1)$ and $(1, 4)$.

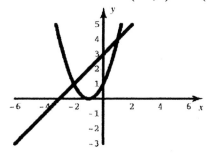

Difficulty: 2   Section: 4

68. If $f(x) = 2x^2 - 3x - 4$, find $\dfrac{f(5) - f(3)}{2}$.

Answer: 13

Difficulty: 1   Section: 5

69. If $h(x) = 3x^2 + 4x - 7$, find $\dfrac{h(6) - h(4)}{4}$.

Answer: 17

Difficulty: 1   Section: 5

70. If $f(x) = 5x^2 + 2x - 10$, find $\dfrac{f(6) - f(3)}{4}$.

    Answer: 35.25

    Difficulty: 1   Section: 5

71. Find the natural domain of the function $f(x) = \dfrac{3x+1}{2x-6}$.

    Answer: All x except $x = 3$

    Difficulty: 1   Section: 5

72. Find the natural domain of the function $f(x) = \dfrac{2x+1}{3x+2}$.

    Answer: All x except $x = -\dfrac{2}{3}$

    Difficulty: 1   Section: 5

73. Find the natural domain of the function $f(x) = \dfrac{x^2+5}{9-x^2}$.

    Answer: All x except $x = \pm 3$

    Difficulty: 1   Section: 5

74. Find the natural domain of the function $f(x) = \dfrac{x+5}{2x^2-8}$.

    Answer: All x except $x = \pm 2$

    Difficulty: 1   Section: 5

75. Find the natural domain of the function $g(x) = \sqrt{5 - 10x}$.

    Answer: All x with $x \leq \dfrac{1}{2}$

    Difficulty: 2   Section: 5

76. Find the natural domain of the function $I(x) = \dfrac{4x}{\sqrt{2x+8}}$.

    Answer: All x with $x > -4$

    Difficulty: 2   Section: 5

77. Find the natural domain of the function $f(x) = \dfrac{1}{\sqrt{x^2+1}}$.

    Answer: All real numbers

Difficulty: 2   Section: 5

78. Find the range of the function $f(x) = \dfrac{1}{x-4}$.

    Answer: All y with $y \neq 0$

    Difficulty: 1   Section: 5

79. Find the range of the function $g(x) = \sqrt{2x - 10}$.

    Answer: All y with $y \geq 0$

    Difficulty: 1   Section: 5

80. Find the range of the function $h(x) = 4 + (x - 7)^2$.

    Answer: All y with $y \geq 4$

    Difficulty: 2   Section: 5

81. Find the range of the function $g(x) = \sqrt{3x + 2}$.

    Answer: All y with $y \geq 0$

    Difficulty: 2   Section: 5

82. Find the domain and range of the function
    $$f(t) = \begin{cases} -1 & \text{for } -3 < t < 0 \\ t & \text{for } t \geq 0 \end{cases}$$

    Answer: The domain is $(-3, \infty)$ and the range is $\{-1\} \cup [0, \infty)$

    Difficulty: 2   Section: 5

83. Find the domain and range of the function
    $$f(\theta) = \begin{cases} 2\sin\theta & \text{for } -\pi < \theta \leq 2\pi \\ \cos\theta & \text{for } 2\pi < \theta < 4\pi \end{cases}$$

    Answer: The domain is $(-\pi, 4\pi)$ and the range is $[-2, 2]$.

    Difficulty: 3   Section: 5

84. Find the domain and range of the function
    $$g(s) = \begin{cases} \sqrt{s} & \text{for } s \geq 4 \\ 0 & \text{for } 0 \leq s < 4 \end{cases}$$

    Answer: The domain is $[0, \infty)$ and the range is $\{0\} \cup [2, \infty)$.

    Difficulty: 2   Section: 5

85. Find the domain and range of the function

$$h(x) = \begin{cases} 2 & \text{for} \quad -1 \leq x < 0 \\ 1 & \text{for} \quad 0 \leq x < 2 \\ 6 - x^2 & \text{for} \quad x \geq 2 \end{cases}$$

Answer: The domain is $[-1, \infty)$ and the range is $(-\infty, 2]$.

Difficulty: 2   Section: 5

86. Determine whether the graph shown is the graph of a function.

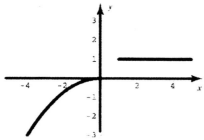

Answer: Yes

Difficulty: 1   Section: 5

87. Determine whether the graph shown is the graph of a function.

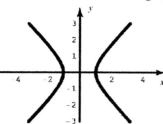

Answer: No

Difficulty: 1   Section: 5

88. Determine whether the graph shown is the graph of a function.

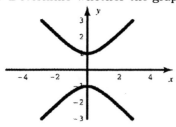

Answer: No

Difficulty: 1   Section: 5

89. Determine whether the graph shown is the graph of a function.

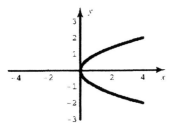

Answer: No

Difficulty: 1   Section: 5

90. Determine whether the graph shown is the graph of a function.

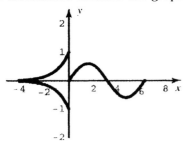

Answer: No

Difficulty: 1   Section: 5

91. Determine whether the graph shown is the graph of a function.

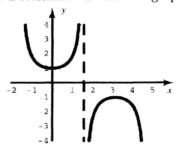

Answer: Yes

Difficulty: 1   Section: 5

92. Determine whether the function $y = \sqrt[3]{x}$ is even, odd or neither.

Answer: Odd

Difficulty: 1   Section: 5

92. Determine whether the function $f(x) = \sqrt{x^2 + 9}$ is even, odd, or neither.

Answer: Even

Difficulty: 2   Section: 5

93. Determine whether the function $f(x) = |2x - 5|$ is even, odd or neither.

Answer: Neither

Difficulty: 2   Section: 5

94. Let $f(x) = \dfrac{1}{x-1}$. Find and simplify $\dfrac{f(x+h) - f(x)}{h}$.

    Answer: $\dfrac{-1}{(x+h-1)(x-1)}$

    Difficulty: 2   Section: 5

95. If $F(x) = \sqrt{x-1}$ and $G(x) = 3x + 2$, give a formula for $(F \circ G)(x)$.

    Answer: $\sqrt{3x+1}$

    Difficulty: 2   Section: 6

96. If $F(x) = 2x + 5$ and $G(x) = x^2 - 3x$, give a formula for $(G \circ F)(x)$.

    Answer: $4x^2 + 14x + 10$

    Difficulty: 2   Section: 6

97. If $F(x) = \sqrt{x^2 + 1}$ and $G(x) = 3x - 2$, give a formula for $(F \circ G)(x)$.

    Answer: $\sqrt{9x^2 - 12x + 5}$

    Difficulty: 2   Section: 6

98. If $F(x) = |x^2 - x|$ and $G(x) = 3x - 1$, give a formula for $(F \circ G)(x)$.

    Answer: $|9x^2 - 9x + 2|$

    Difficulty: 2   Section: 6

99. If $F(x) = \dfrac{x}{x+2}$ and $G(x) = \dfrac{3x}{x+1}$, give a formula for $(F \circ G)(x)$.

    Answer: $\dfrac{3x}{5x+2}$

    Difficulty: 3   Section: 6

100. If $F(x) = \dfrac{x}{x-1}$ and $G(x) = 2x + 1$, give a formula for $(F \circ G)(x)$.

    Answer: $\dfrac{2x+1}{2x}$

    Difficulty: 2   Section: 6

101. If $F(x) = 3x + 5$ and $G(x) = \dfrac{x}{x+2}$, give a formula for $(F \circ G)(x)$.

    Answer: $\dfrac{8x+10}{x+2}$

Difficulty: 2   Section: 6

102. If $F(x) = \dfrac{2x}{x+3}$ and $G(x) = \dfrac{x}{x-1}$, give a formula for $(F \circ G)(x)$.

Answer: $\dfrac{2x}{4x-3}$

Difficulty: 3   Section: 6

103. Let $f(x) = 2x - 3$ and $g(x) = x^2 - 3x - 28$. Write a formula for $(f+g)(x)$ and $(f \cdot g)(x)$ and state the domain of each.

Answer: $(f+g)(x) = x^2 - x - 31$ and $(f \cdot g)(x) = 2x^3 - 9x^2 - 47x + 84$.
Domain consists of all real numbers in both cases.

Difficulty: 2   Section: 6

104. Let $f(x) = \sqrt{2x-3}$ and $g(x) = \sqrt{3x+1}$. Write a formula for $(f+g)(x)$ and $(f \cdot g)(x)$ and state the domain of each.

Answer: $(f+g)(x) = \sqrt{2x-3} + \sqrt{3x+1}$ and $(f \cdot g)(x) = \sqrt{6x^2 - 7x - 3}$.
Domain consists of $\left\{x : x \geq \dfrac{3}{2}\right\}$ in both cases.

Difficulty: 2   Section: 6

105. Let $f(x) = x - 1$ and $g(x) = x^2 - 1$. Write a formula for $(f \cdot g)(x)$ and $(f/g)(x)$ and state the domain of each.

Answer: $(f \cdot g)(x) = x^3 - x^2 - x + 1$ and $(f/g)(x) = \dfrac{1}{x+1}$
Domain consists of all reals in the first case and all reals except $x = \pm 1$ in the second case.

Difficulty: 2   Section: 6

106. Graph the following functions of the same coordinate axes:
$y = \sqrt{x}$, $y = \sqrt{x+2}$, $y = 3 + \sqrt{x}$, $y - 3 = \sqrt{x+2}$

Answer:

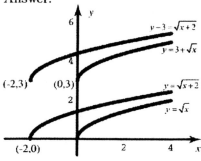

Difficulty: 2   Section: 6

107. Graph the following functions of the same coordinate axes:

$y = x^2$, $y = 2x^2$, $y = -x^2$, $y = -\dfrac{1}{2}x^2$

Answer:

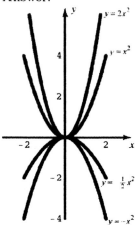

Difficulty: 2   Section: 6

108. Convert 75° to radian measure.
   Answer: $\dfrac{5\pi}{12}$
   Difficulty: 1   Section: 7

109. Convert 20° to radian measure.
   Answer: $\dfrac{\pi}{9}$
   Difficulty: 1   Section: 7

110. Convert 40° to radian measure.
   Answer: $\dfrac{2\pi}{9}$
   Difficulty: 1   Section: 7

111. Convert 220° to radian measure.
   Answer: $\dfrac{11\pi}{9}$
   Difficulty: 1   Section: 7

112. Convert 320° to radian measure.
   Answer: $\dfrac{16\pi}{9}$
   Difficulty: 1   Section: 7

113. Convert 14° to radian measure.

Answer: $\dfrac{7\pi}{90}$

Difficulty: 1   Section: 7

114. Convert $100°$ to radian measure.

    Answer: $\dfrac{5\pi}{9}$

    Difficulty: 1   Section: 7

115. Convert $-\dfrac{5\pi}{3}$ to degree measure.

    Answer: $-300°$

    Difficulty: 1   Section: 7

116. Convert $\dfrac{5\pi}{2}$ to degree measure.

    Answer: $450°$

    Difficulty: 1   Section: 7

117. Convert $-\dfrac{2\pi}{3}$ to degree measure.

    Answer: $-120°$

    Difficulty: 1   Section: 7

118. Convert $\dfrac{4\pi}{9}$ to degree measure.

    Answer: $80°$

    Difficulty: 1   Section: 7

119. Convert $\dfrac{5\pi}{18}$ to degree measure.

    Answer: $50°$

    Difficulty: 1   Section: 7

120. Convert $-\dfrac{11\pi}{9}$ to degree measure.

    Answer: $-220°$

    Difficulty: 1   Section: 7

121. Convert $-\dfrac{4\pi}{3}$ to degree measure.

    Answer: $-240°$

    Difficulty: 1   Section: 7

122. Use the right triangle to determine $\sin\theta$, $\cos\theta$, and $\tan\theta$.

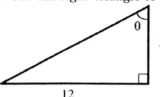

Answer: $\sin\theta = \dfrac{12}{13}$, $\cos\theta = \dfrac{5}{13}$, $\tan\theta = \dfrac{12}{5}$

Difficulty: 2   Section: 7

123. Use the right triangle to determine $\sin\theta$, $\cos\theta$, and $\tan\theta$.

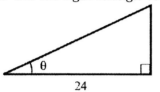

Answer: $\sin\theta = \dfrac{7}{25}$, $\cos\theta = \dfrac{24}{25}$, $\tan\theta = \dfrac{7}{24}$

Difficulty: 2   Section: 7

124. Use the right triangle to determine $\sin\theta$, $\cos\theta$, and $\tan\theta$.

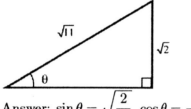

Answer: $\sin\theta = \sqrt{\dfrac{2}{11}}$, $\cos\theta = \dfrac{3}{\sqrt{11}}$, $\tan\theta = \dfrac{\sqrt{2}}{3}$

Difficulty: 2   Section: 7

125. Use the right triangle to determine $\sin\theta$, $\cos\theta$, and $\tan\theta$.

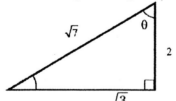

Answer: $\sin\theta = \sqrt{\dfrac{3}{7}}$, $\cos\theta = \dfrac{2}{\sqrt{7}}$, $\tan\theta = \dfrac{\sqrt{3}}{2}$

Difficulty: 2   Section: 7

126. Find the value of $\sin^2\left(\dfrac{\pi}{5}\right) + \cos^2\left(\dfrac{\pi}{5}\right)$.

Answer: 1

Difficulty: 1   Section: 7

127. Evaluate $\sin t$, $\cos t$, and $\tan t$ for $t = \dfrac{\pi}{3}$.

   Answer: $\dfrac{\sqrt{3}}{2}, \dfrac{1}{2}, \sqrt{3}$

   Difficulty: 2   Section: 7

128. Evaluate $\sin t$, $\cos t$, and $\tan t$ for $t = \pi$.

   Answer: $0, -1, 0$

   Difficulty: 2   Section: 7

129. Evaluate $\sin t$, $\cos t$, and $\tan t$ for $t = -\dfrac{\pi}{4}$.

   Answer: $-\dfrac{\sqrt{2}}{2}, \dfrac{\sqrt{2}}{2}, -1$

   Difficulty: 2   Section: 7

130. Find the value of $\csc t$ if $\sin t = -\dfrac{4}{5}$.

   Answer: $-\dfrac{5}{4}$

   Difficulty: 1   Section: 7

131. Verify that $\sin t = \sqrt{1 - \cos^2 t}$ is not an identity.

   Answer: Varies. $t = \dfrac{3\pi}{2}$ is a counterexample.

   Difficulty: 2   Section: 7

132. Find the signs of $\sin t$ and $\cos t$ if $t = 7$.

   Answer: Both are positive.

   Difficulty: 2   Section: 7

133. Find the signs of $\sin t$ and $\cos t$ if $t = -\dfrac{7\pi}{6}$.

   Answer: $\sin t > 0$ and $\cos t < 0$.

   Difficulty: 2   Section: 7

134. Sketch the graphs of $y = \sin t$, $y = 2\sin t$, and $y = -2\sin t$ on $[0, 2\pi]$

   Answer:

   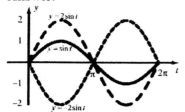

Difficulty: 2   Section: 7

135. Sketch the graph of $2\cos\frac{1}{2}t$ on $[0, 2\pi]$.

Answer:

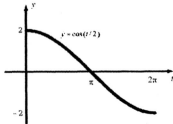

Difficulty: 3   Section: 7

1 Limits

1. Let $f(x) = 3 - x^3$. Evaluate the limit $\lim_{x \to -2} f(x)$.

   Answer: 11

   Difficulty: 1   Section: 1

2. Let $g(x) = \dfrac{x-1}{x-5}$. Evaluate the limit $\lim_{x \to 5} g(x)$.

   Answer: The limit does not exist.

   Difficulty: 1   Section: 1

3. Let $s(x) = |x - 5|$. Evaluate the limit $\lim_{x \to -5} s(x)$.

   Answer: 10

   Difficulty: 1   Section: 1

4. Let $g(x) = \dfrac{x^2 - 4}{x + 2}$. Evaluate the limit $\lim_{x \to -2} g(x)$.

   Answer: -4

   Difficulty: 2   Section: 1

5. Let $t(x) = \cos 3x$. Evaluate the limit $\lim_{x \to \pi/3} t(x)$.

   Answer: -1

   Difficulty: 2   Section: 1

6. Let $f(x) = \sqrt{6 - x}$. Evaluate the limit $\lim_{x \to 3} f(x)$.

   Answer: $\sqrt{3}$

   Difficulty: 1   Section: 1

7. Let $f(x) = \dfrac{2 + x}{\cos x}$. Evaluate the limit $\lim_{x \to \pi/2} f(x)$.

   Answer: The limit does not exist.

   Difficulty: 2   Section: 1

8. Let $g(x) = \dfrac{1}{|x - 1| + 1}$. Evaluate the limit $\lim_{x \to -1} g(x)$.

   Answer: $\dfrac{1}{3}$

   Difficulty: 2   Section: 1

9. Let $f(x) = \begin{cases} x & \text{for } 0 \leq x \leq 1 \\ \sqrt{x} & \text{for } x > 1 \end{cases}$. Evaluate the limit $\lim_{x \to 1} f(x)$.

   Answer: 1

   Difficulty: 2   Section: 1

10. Let $f(x) = \begin{cases} \sqrt{x-1} & \text{for } x \geq 1 \\ x & \text{for } x < 1 \end{cases}$. Evaluate the limit $\lim_{x \to 1} f(x)$.

    Answer: The limit does not exist.

    Difficulty: 2   Section: 1

11. Let $g(x) = \dfrac{|x|}{x}$. Evaluate the limit $\lim_{x \to -3} g(x)$.

    Answer: -1

    Difficulty: 2   Section: 1

12. Let $g(x) = \dfrac{|x|}{x}$. Evaluate the limit $\lim_{x \to 0} g(x)$.

    Answer: The limit does not exist.

    Difficulty: 2   Section: 1

13. Evaluate $\lim_{x \to -5^+} \sqrt{x+5}$.

    Answer: 0

    Difficulty: 2   Section: 1

14. Evaluate $\lim_{x \to -5^-} \sqrt{x+5}$.

    Answer: The left limit does not exist.

    Difficulty: 2   Section: 1

15. Evaluate $\lim_{x \to 0^+} \dfrac{|x|}{x}$.

    Answer: 1

    Difficulty: 2   Section: 1

16. Evaluate $\lim_{x \to 0^-} \dfrac{|x|}{x}$.

    Answer: -1

    Difficulty: 2   Section: 1

17. Evaluate $\lim_{x \to 0^+} \dfrac{1 - \sqrt{x}}{1 - x}$.

Answer: 1

Difficulty: 2   Section: 1

18. Evaluate $\lim_{x\to 0^-} \dfrac{1-\sqrt{x}}{1-x}$.

    Answer: The left limit does not exist.

    Difficulty: 2   Section: 1

19. Use a calculator to evaluate $\lim_{x\to 0} \dfrac{\sin 2x}{\sin 3x}$.

    Answer: $\dfrac{2}{3}$

    Difficulty: 2   Section: 1

20. Let $\llbracket x \rrbracket$ denote the greatest integer less than or equal to x. Evaluate $\lim_{x\to -2^+} \llbracket x \rrbracket$.

    Answer: -2

    Difficulty: 2   Section: 1

21. Let $\llbracket x \rrbracket$ denote the greatest integer less than or equal to x. Evaluate $\lim_{x\to -2^-} \llbracket x \rrbracket$.

    Answer: -3

    Difficulty: 2   Section: 1

22. Let $\llbracket x \rrbracket$ denote the greatest integer less than or equal to x. Evaluate $\lim_{x\to -3^-} \dfrac{\llbracket x \rrbracket}{x}$.

    Answer: $\dfrac{4}{3}$

    Difficulty: 2   Section: 1

23. Give an ε, δ proof that $\lim_{x\to 3} x^2 = 9$.

    Answer:
    Choose $\delta = \min\left(1, \dfrac{\varepsilon}{7}\right)$. If $0 < |x-3| < \delta$ then
    $|x^2 - 9| = |x+3||x-3| < 7|x-3| < 7\left(\dfrac{\varepsilon}{7}\right) = \varepsilon$.

    Difficulty: 2   Section: 2

24. Give an ε, δ proof that $\lim_{x\to 3} x^3 = 27$.

    Answer:
    Choose $\delta = \min\left(1, \dfrac{\varepsilon}{37}\right)$. If $0 < |x-3| < \delta$ then
    $|x^3 - 27| = |x^2 + 3x + 9||x-3| < 37|x-3| < 37\left(\dfrac{\varepsilon}{37}\right) = \varepsilon$.

    Difficulty: 2   Section: 2

25. Give an ε, δ proof that $\lim_{x \to 2} (3x + 2) = 8$.

   Answer:
   Choose $\delta = \dfrac{\varepsilon}{3}$. If $0 < |x - 2| < \delta$ then
   $$|(3x + 2) - 8| = |3x - 6| = 3|x - 2| < 3\left(\dfrac{\varepsilon}{3}\right) = \varepsilon.$$

   Difficulty: 2   Section: 2

26. Give an ε, δ proof that $\lim_{x \to 3} (2 - 5x) = -13$.

   Answer:
   Choose $\delta = \dfrac{\varepsilon}{5}$. If $0 < |x - 3| < \delta$ then
   $$|(2 - 5x) - (-13)| = |15 - 5x| = 5|3 - x| = 5|x - 3| < 5\left(\dfrac{\varepsilon}{5}\right) = \varepsilon.$$

   Difficulty: 2   Section: 2

27. Let $[\![x]\!]$ denote the greatest integer less than or equal to x. Give an ε, δ argument that $\lim_{x \to 0} [\![x]\!]$ does not exist.

   Answer:
   Suppose that $\lim_{x \to 0} [\![x]\!] = L$. Let $\varepsilon = \dfrac{1}{2}$. Then there is a $\delta > 0$ such that $|[\![x]\!] - L| < \dfrac{1}{2}$ whenever $0 < |x| < \delta$. If $L \geq 0$, $x = -\dfrac{\delta}{2}$ violates this inequality and if $L < 0$, $x = \dfrac{\delta}{2}$ violates the inequality.

   Difficulty: 3   Section: 2

28. Let $\varepsilon > 0$ be given. Find the largest δ that will "work" for $\lim_{x \to 4} 3x = 12$.

   Answer: $\dfrac{\varepsilon}{3}$

   Difficulty: 1   Section: 2

29. Let $\varepsilon > 0$ be given. Find the largest δ that will "work" for $\lim_{x \to 6} \dfrac{1}{3}x = 2$.

   Answer: 3ε

   Difficulty: 1   Section: 2

30. Evaluate $\lim_{x \to -1} \dfrac{81 + x^4}{9 + x}$.

   Answer: $\dfrac{41}{4}$

   Difficulty: 2   Section: 3

31. Evaluate $\lim_{x \to -9} \dfrac{81 + x^4}{9 + x}$.

   Answer: The limit does not exist.

Difficulty: 2   Section: 3

32. Let $f(x) = \begin{cases} 2x + c & \text{for } x < 2 \\ (x-5)^2 & \text{for } x \geq 2 \end{cases}$.
    What is the value of c if $\lim_{x \to 2} f(x) = 9$?

    Answer: 5

    Difficulty: 2   Section: 3

33. Let $f(x) = 2x - 3$. Determine $\lim_{h \to 0} \dfrac{f(2+h) - f(2)}{h}$.

    Answer: 2

    Difficulty: 3   Section: 3

34. Let $f(x) = \sqrt{x}$. Find $\lim_{h \to 0} \dfrac{f(9+h) - f(9)}{h}$.

    Answer: $\dfrac{1}{6}$

    Difficulty: 3   Section: 3

35. Let $\llbracket x \rrbracket$ denote the greatest integer less than or equal to x. Evaluate $\lim_{x \to -1/2} \llbracket x \rrbracket$.

    Answer: -1

    Difficulty: 2   Section: 3

36. If $\lim_{x \to a} f(x) = M$ and $\lim_{x \to a} g(x) = N$, find
    $$\lim_{x \to a} \frac{2[f(x)]^2 - g(x)}{1 + f(x)g(x)} \text{ if } MN \neq -1.$$

    Answer: $\dfrac{2M^2 - N}{1 + MN}$

    Difficulty: 2   Section: 3

37. Let $f(x) = \begin{cases} x^3 - 1 & \text{if } x \leq 1 \\ x^2 & \text{if } x > 1 \end{cases}$. Evaluate $\lim_{x \to 1^-} f(x)$.

    Answer: 0

    Difficulty: 2   Section: 3

38. Evaluate $\lim_{x \to 0^-} (2 - x) \cos x$.

    Answer: 2

    Difficulty: 2   Section: 3

39. Evaluate: $\lim_{x \to 4^+} \sqrt{x^2 - 3x}$.

Answer: 2

Difficulty: 2   Section: 3

40. Evaluate $\lim_{x \to 3^-} \sqrt{x^2 - 3x}$.

    Answer: The left limit does not exist.

    Difficulty: 2   Section: 3

41. Find the value of k if $\lim_{x \to 0^-} \left(k + \dfrac{|x|}{x}\right) = 7$.

    Answer: 8

    Difficulty: 2   Section: 3

42. Evaluate $\lim_{x \to -1} \dfrac{x^4 + x^3 + x^2 + x + 1}{(1-x)^2}$.

    Answer: $\dfrac{1}{4}$

    Difficulty: 2   Section: 3

43. Evaluate $\lim_{x \to 0} x \cot x$.

    Answer: 1

    Difficulty: 2   Section: 3

44. Evaluate $\lim_{x \to 0} \dfrac{\sin 4x}{\sin 3x}$.

    Answer: $\dfrac{4}{3}$

    Difficulty: 2   Section: 4

45. Evaluate $\lim_{x \to \pi/6} \dfrac{\sin 3x}{\cos 2x}$.

    Answer: 2

    Difficulty: 2   Section: 4

46. Evaluate $\lim_{x \to 0} \dfrac{x + \sin(x^2)}{x}$.

    Answer: 1

    Difficulty: 2   Section: 4

47. Evaluate $\lim_{x \to 0} \dfrac{x}{1 - \cos x}$.

    Answer: The limit does not exist.

Difficulty: 3   Section: 4

48. Evaluate $\lim\limits_{x \to 0} \dfrac{x}{\tan 3x}$.

    Answer: $\dfrac{1}{3}$

    Difficulty: 2   Section: 4

49. Evaluate $\lim\limits_{x \to 0} \dfrac{1 - \cos x}{\pi x}$.

    Answer: 0

    Difficulty: 2   Section: 4

50. Evaluate $\lim\limits_{x \to 0^+} \dfrac{x^2 - 5x}{\tan x}$.

    Answer: -5

    Difficulty: 2   Section: 4

51. Let $f(x) = \begin{cases} \dfrac{\sin x}{2x} & \text{for } x > 0 \\ \cos 2x & \text{for } x \leq 0 \end{cases}$. Evaluate $\lim\limits_{x \to 0^+} f(x)$.

    Answer: $\dfrac{1}{2}$

    Difficulty: 2   Section: 4

52. Let $f(x) = \begin{cases} \dfrac{\sin x}{2x} & \text{for } x > 0 \\ \cos 2x & \text{for } x \leq 0 \end{cases}$. Evaluate $\lim\limits_{x \to 0^-} f(x)$.

    Answer: 1

    Difficulty: 2   Section: 4

53. Evaluate $\lim\limits_{x \to \pi} [\tan x - \pi]$.

    Answer: $-\pi$

    Difficulty: 2   Section: 4

54. Evaluate $\lim\limits_{x \to -\pi/4} \tan\left(x - \dfrac{\pi}{2}\right)$.

    Answer: 1

    Difficulty: 2   Section: 4

55. Evaluate $\lim\limits_{x \to \pi/2} \dfrac{\sin x}{x}$

Answer: $\dfrac{2}{\pi}$

Difficulty: 1   Section: 4

56. Evaluate $\lim\limits_{x \to \infty} \dfrac{7x^3 - 4x + 15}{3x^3 - 14x}$.

   Answer: $\dfrac{7}{3}$

   Difficulty: 1   Section: 5

57. Evaluate $\lim\limits_{x \to \infty} \dfrac{2x + 3}{2 - 3x}$.

   Answer: $-\dfrac{2}{3}$

   Difficulty: 1   Section: 5

58. Evaluate $\lim\limits_{x \to -\infty} \dfrac{4x^2 - 7x - 5}{13x + 6}$.

   Answer: $-\infty$

   Difficulty: 2   Section: 5

59. Evaluate $\lim\limits_{x \to 4^+} \left(\dfrac{-2}{x - 4} + 2 \right)$.

   Answer: $-\infty$

   Difficulty: 2   Section: 5

60. Evaluate $\lim\limits_{x \to 4^-} \left(\dfrac{-2}{x - 4} + 2 \right)$.

   Answer: ∞

   Difficulty: 2   Section: 5

61. Evaluate $\lim\limits_{x \to 1^-} \left[\dfrac{x - 4}{-2(x^2 - 8x + 7)} \right]$.

   Answer: ∞

   Difficulty: 2   Section: 5

62. Evaluate $\lim\limits_{x \to 7^+} \left[\dfrac{x - 4}{-2(x^2 - 8x + 7)} \right]$.

   Answer: $-\infty$

   Difficulty: 2   Section: 5

63. Determine all points at which $f(x) = \dfrac{1}{x^2 + 1}$ is continuous.

Answer: All real numbers

Difficulty: 2   Section: 6

64. Determine all points at which $f(x) = \begin{cases} \dfrac{\sin 2x}{2x} & \text{for } x \neq 0 \\ 2 & \text{for } x = 0 \end{cases}$ is continuous.

Answer: All real numbers except 0

Difficulty: 3   Section: 6

65. Determine all points at which $f(x) = \sqrt{x-5}$ is continuous.

Answer: $x \geq 5$

Difficulty: 2   Section: 6

66. Determine all points at which $f(x) = \dfrac{x-2}{|x-2|}$ is continuous.

Answer: Continuous at every point in the domain. Note that 2 is not in the domain of f.

Difficulty: 2   Section: 6

67. Determine all points at which $f(x) = \begin{cases} \sqrt{x} & \text{for } x \geq 0 \\ x^2 - x & \text{for } x < 0 \end{cases}$ is continuous.

Answer: All real numbers

Difficulty: 2   Section: 6

68. Determine all points at which $f(x) = \begin{cases} \dfrac{2x^2 + 5x}{x} & \text{for } x \neq 0 \\ 5 & \text{for } x = 0 \end{cases}$ is continuous.

Answer: All real numbers

Difficulty: 2   Section: 6

69. Determine all points at which $f(x) = \begin{cases} \dfrac{x^2 - 9}{x - 3} & \text{for } x \neq 3 \\ 3 & \text{for } x = 3 \end{cases}$ is continuous.

Answer: All real numbers except 3.

Difficulty: 2   Section: 6

70. Let $f(x) = \begin{cases} 1 & \text{if } x \text{ is an integer} \\ 0 & \text{otherwise} \end{cases}$. Determine all points at which f is continuous.

Answer: Continuous at all real numbers that are not integers.

Difficulty: 2   Section: 6

71. The function $f(x) = \dfrac{x^3 - 1}{x - 1}$ is continuous at every point of the domain. Define $f(1)$ so that

f is continuous at every real number.

Answer: $f(1) = 3$

Difficulty: 2   Section: 6

72. The function $g(x) = \dfrac{\sin x}{2x}$ is continuous at every point of its domain. Define $g(0)$ so that g is continuous at every real number.

Answer: $g(0) = \dfrac{1}{2}$

Difficulty: 3   Section: 6

73. Let $f(x) = x^3 + x - 1$. Show that $f(x) = 0$ for some x in the interval $[0, 1]$.

Answer: $f(0) = -1$, $f(1) = 1$. Use the Intermediate Value Theorem.

Difficulty: 2   Section: 6

74. Let $f(x) = 2\sin(x - 1) + 1$. Show that $f(x) = 0$ for some x in the interval $\left[0, \dfrac{\pi}{2}\right]$.

Answer: $f(0) < 0$, $f\left(\dfrac{\pi}{2}\right) > 0$. Use the Intermediate Value Theorem.

Difficulty: 2   Section: 6

Problems 75–80 refer to the figure below.

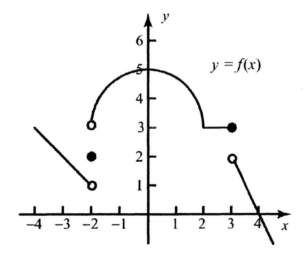

75. What is $\lim\limits_{x \to -2^-} f(x)$?

Answer: 1

Difficulty: 1   Section: 1

76. What is $\lim\limits_{x \to -2} f(x)$?

Answer: This limit does not exist.

Difficulty: 1   Section: 1

77. What is $\lim\limits_{x \to 2^+} f(x)$?

    Answer: 3

    Difficulty: 1   Section: 1

78. Is f continuous at $x = 2$?

    Answer: Yes

    Difficulty: 1   Section: 1

79. Is f continuous on the interval $(-2, 3)$?

    Answer: Yes

    Difficulty: 1   Section: 6

80. Is f continuous on the interval $[-2, 3]$?

    Answer: No

    Difficulty: 1   Section: 6

Problems 81-86 refer to the figure below.

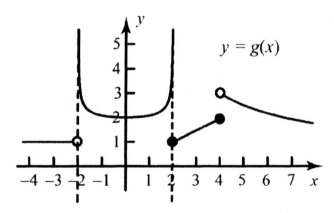

81. What is $\lim\limits_{x \to -2^-} g(x)$?

    Answer: 1

    Difficulty: 1   Section: 1

82. What is $\lim\limits_{x \to -2} g(x)$?

    Answer: This limit does not exist.

    Difficulty: 1   Section: 1

83. What is $\lim_{x \to 2^+} g(x)$?

   Answer: 1

   Difficulty: 1   Section: 1

84. What is $\lim_{x \to 2^-} g(x)$?

   Answer: ∞

   Difficulty: 1   Section: 1

85. Is g continuous on the interval $(-2, 2)$?

   Answer: Yes

   Difficulty: 1   Section: 6

86. Is g continuous on the interval $[-2, 2]$?

   Answer: No

   Difficulty: 1   Section: 6

2 The Derivative

1. For the position function $s = t^2 - 1$, calculate the instantaneous velocity at time $t_0 = 4$.

   Answer: 8

   Difficulty: 1   Section: 1

2. For the position function $s = 7t - 2$, calculate the instantaneous velocity at time $t_0 = 9$.

   Answer: 7

   Difficulty: 1   Section: 1

3. Find the equation of the tangent line to the graph of $y = \dfrac{3}{x}$ at the point where $x = -2$.

   Answer: $y = -\dfrac{3}{4}(x+2) - \dfrac{3}{2}$

   Difficulty: 2   Section: 1

4. Find the equation of the tangent line to the graph of $y = \sqrt{x}$ at the point where $x = 9$.

   Answer: $y = \dfrac{1}{6}(x-9) + 3$

   Difficulty: 2   Section: 1

5. Let $f(x) = -5x^2$. Use the limit definition of derivative to compute $f'(x)$ and then find $f'(-5)$.

   Answer: $f'(x) = \lim\limits_{h \to 0} \dfrac{f(x+h) - f(x)}{h} = -10x; 50.$

   Difficulty: 2   Section: 2

6. Let $f(x) = -\dfrac{1}{x}$. Use the limit definition of derivative to compute $f'(x)$ and then find $f'(-7)$.

   Answer: $f'(x) = \lim\limits_{h \to 0} \dfrac{f(x+h) - f(x)}{h} = \dfrac{1}{x^2}; \dfrac{1}{49}$

   Difficulty: 2   Section: 2

7. Use the limit definition of derivative to find $p'(q)$ if $p = 1 - q^2$.

   Answer: $p'(q) = \lim\limits_{h \to 0} \dfrac{\left[1 - (q+h)^2\right] - \left[1 - q^2\right]}{h} = -2q.$

   Difficulty: 2   Section: 2

8. Use the limit definition of derivative to find $v'(t)$ if $v = t - t^3$.

   Answer: $v'(t) = \lim\limits_{h \to 0} \dfrac{\left[(t+h) - (t+h)^3\right] - \left[t - t^3\right]}{h} = 1 - 3t^2.$

Difficulty: 2   Section: 2

9. Let $f(x) = \begin{cases} 0 & \text{if } x \leq 0 \\ 2x+1 & \text{if } x > 0 \end{cases}$. Prove that $f(x)$ is not differentiable at $x = 0$.

   Answer: $\lim\limits_{h \to 0} \dfrac{f(0+h) - f(0)}{h} = 2$; $\lim\limits_{h \to 0^-} \dfrac{f(0+h) - f(0)}{h} = 0$.

   Difficulty: 2   Section: 2

10. Examine the graph of $f(x) = |5x + 12|$ and determine the points (if any) where $f'(x)$ does not exist.

    Answer: $-\dfrac{12}{5}$

    Difficulty: 2   Section: 2

11. Examine the graph of $f(x) = [x]$, where $[x] = $ greatest integer $\leq x$, and determine the points (if any) where $f'(x)$ does not exist.

    Answer: The derivative does not exist at any integer value of x.

    Difficulty: 2   Section: 2

12. Give an example of a function $f(x)$, defined for all real numbers, such that $f'(x) = 0$ for all x.

    Answer: $f(x) = c$ for any constant c.

    Difficulty: 2   Section: 2

13. Find $D_x y$ if $y = 5x^5 - 2x^{-1}$.

    Answer: $25x^4 + 2x^{-2}$

    Difficulty: 1   Section: 3

14. Find $D_x y$ if $y = x^4 - x - x^{-2}$.

    Answer: $4x^3 - 1 + 2x^{-3}$

    Difficulty: 1   Section: 3

15. Find $D_x y$ if $y = \sqrt[3]{x} + x - 10$.

    Answer: $\dfrac{1}{3} x^{-2/3} + 1$

    Difficulty: 2   Section: 3

16. Let $f(t) = (2t^5 - t)(t^3 + 1)$. Compute $f'(t)$ and simplify.

    Answer: $16t^7 + 10t^4 - 4t^3 - 1$

    Difficulty: 2   Section: 3

17. Let $y = (x^2 + 2x - 5)(x^3 - 1)$. Compute $D_x y$ using the product rule and simplify.

    Answer: $5x^4 + 8x^3 - 15x^2 - 2x - 2$

    Difficulty: 2   Section: 3

18. Let $f(x) = \dfrac{x^2 - 1}{x^2 + 1}$. Find $f'(x)$.

    Answer: $\dfrac{4x}{(x^2 + 1)^2}$

    Difficulty: 2   Section: 3

19. Let $y = \dfrac{1 + x}{1 - x} - 4x^3$. Find $D_x y$.

    Answer: $\dfrac{2}{(1 - x)^2} - 12x^2$

    Difficulty: 2   Section: 3

20. Let $f(x) = \dfrac{x^2 + 1}{x - 3}$. Compute $f'(x)$ and simplify.

    Answer: $\dfrac{x^2 - 6x - 1}{(x - 3)^2}$

    Difficulty: 2   Section: 3

21. Let $g(x) = \dfrac{x + 1}{2 - x^2}$. Compute $g'(x)$ and simplify.

    Answer: $\dfrac{x^2 + 2x + 2}{(2 - x^2)^2}$

    Difficulty: 2   Section: 3

22. Let $f(t) = \dfrac{t^3 - 1}{t^2 + 3t - 2}$. Find $f'(t)$ and simplify.

    Answer: $\dfrac{t^4 + 6t^3 - 6t^2 + 2t + 3}{(t^2 + 3t - 2)^2}$

    Difficulty: 2   Section: 3

23. Find an equation of the tangent line to the graph of the function $y = 1 + x^{-1} + x^{-1} + x^{-2} + x^{-3}$ at the point $(1, 4)$.

    Answer: $y + 6x - 10 = 0$

    Difficulty: 2   Section: 3

24. Find an equation of the tangent line to the graph of the function $f(x) = -3 + \dfrac{1}{x} - \dfrac{1}{x^2} + \dfrac{1}{x^3}$

at the point $(1, -2)$.

Answer: $y + 2x = 0$

Difficulty: 2   Section: 3

25. Find an equation of the tangent line to $f(x) = \sqrt{x} + 5$ at the point $(9, 8)$.

Answer: $x - 6y + 39 = 0$

Difficulty: 2   Section: 3

26. Find an equation of the tangent line to $f(x) = \dfrac{x^2 + 3}{1 - x}$ at the point $(0, 3)$.

Answer: $3x - y + 3 = 0$

Difficulty: 2   Section: 3

27. What is the point on the graph of $y = -2x^2 - 9x - 5$ at which the tangent line is perpendicular to the line $y = x - 10$?

Answer: $(-2, 5)$

Difficulty: 3   Section: 3

28. If $f(x) = \dfrac{1}{x} + \cos x$, find the value of $f'(x)$.

Answer: $-\dfrac{1}{x^2} - \sin(x)$

Difficulty: 1   Section: 4

29. Let $s = -5t \csc(t)$. Find $D_t s$.

Answer: $5 \csc(t) [t \cot(t) - 1]$

Difficulty: 2   Section: 4

30. Let $f(x) = 3 \tan(x) - 2 \sec(x) + 3x^2$. Find $f'(x)$.

Answer: $3 \sec^2(x) - 2 \sec(x) \tan(x) + 6x$

Difficulty: 2   Section: 4

31. Let $y = x^2 \cos(x)$. Find $D_x y$.

Answer: $2x \cos x - x^2 \sin(x)$

Difficulty: 2   Section: 4

32. Let $g(k) = (k^2 + 2k)[1 - \cos(k)]$. Find $g'(k)$.

Answer: $(k^2 + 2k)[\sin(k)] + 2[1 - \cos(k)](k + 1)$

Difficulty: 2   Section: 4

33. Let $f(x) = \dfrac{\sin(x) + \cos(x)}{x^2}$. Find $f'(x)$ and simplify.

    Answer: $\dfrac{x\cos(x) - x\sin(x) - 2\sin(x) - 2\cos(x)}{x^3}$

    Difficulty: 2   Section: 4

34. Find the equation of the tangent line to the curve $y = \cot(x)$ at the point where $x = \dfrac{\pi}{4}$.

    Answer: $y + 2x - 1 - \dfrac{\pi}{2} = 0$

    Difficulty: 2   Section: 4

35. Find the equation of the tangent line to the graph of $y = 2\tan(x)$ at the point where $x = \dfrac{\pi}{3}$.

    Answer: $y - 2\sqrt{3} = 8\left(x - \dfrac{\pi}{3}\right)$

    Difficulty: 2   Section: 4

36. Find the equation of the tangent line to the graph of $y = \sqrt{3}\sec(x) - 2$ at the point where $x = \dfrac{\pi}{6}$.

    Answer: $y = \dfrac{2\sqrt{3}}{3}\left(x - \dfrac{\pi}{6}\right)$

    Difficulty: 2   Section: 4

37. Let $y = (x+1)^5$. Find $D_x y$.

    Answer: $5(x+1)^4$

    Difficulty: 1   Section: 5

38. Let $f(x) = (x^2 - x + 5)^4$. Find $f'(x)$.

    Answer: $4(2x - 1)(x^2 - x + 5)^3$

    Difficulty: 2   Section: 5

39. Let $g(x) = (1 - 3x)^{-2}$. Find $g'(x)$.

    Answer: $6(1 - 3x)^{-3}$

    Difficulty: 2   Section: 5

40. Let $g(x) = \left(x^2 + \dfrac{1}{x^2}\right)^3$. Find $g'(x)$.

    Answer: $6\left(x^2 + \dfrac{1}{x^2}\right)^2\left(x - \dfrac{1}{x^3}\right)$

    Difficulty: 2   Section: 5

41. Let $f(x) = \left(\dfrac{x}{x^2+3}\right)^5$. Find $f'(x)$ and simplify.

    Answer: $\dfrac{5x^4(3-x^2)}{(x^2+3)^6}$

    Difficulty: 2   Section: 5

42. Let $f(x) = \left(\dfrac{1+x}{1-x}\right)^4$. Find $f'(x)$ and simplify.

    Answer: $\dfrac{8(1+x)^3}{(1-x)^5}$

    Difficulty: 2   Section: 5

43. Let $f(x) = \dfrac{1}{x} + \cos 2x$. Find $f'(x)$.

    Answer: $-\left(\dfrac{1}{x^2} + 2\sin 2x\right)$

    Difficulty: 2   Section: 5

44. Let $x = t^4 - \sec^2(t)$. Find $D_t x$.

    Answer: $4t^3 - 2\sec^2 t \tan t$

    Difficulty: 2   Section: 5

45. Let $y = \cos(\sin 2x)$. Find $D_x y$.

    Answer: $-2\sin(\sin 2x)\cos 2x$

    Difficulty: 2   Section: 5

46. Let $y = x\cos(\sin x)$. Find $D_x y$.

    Answer: $\cos(\sin x) - x\sin(\sin x)\cos x$

    Difficulty: 2   Section: 5

47. If the equation of motion of a particle is $x = (2t^2 - 7t)^3$, what is the velocity of the particle when $t = 1$?

    Answer: -225

    Difficulty: 2   Section: 5

48. Find an equation of the tangent line to the graph of $y = (3x^2 + 5x - 1)^7$ at the point where $x = 0$.

    Answer: $y - 35x + 1 = 0$

    Difficulty: 2   Section: 5

49. Find the slope of the line tangent to the graph of $y = x^2 \cos 2x$ at the point where $x = \dfrac{\pi}{2}$.

   Answer: $-\pi$

   Difficulty: 2   Section: 5

50. Making the usual assumptions, write a formula for the derivative of the function $\sin^4 [f(x)]$.

   Answer: $4 \sin^3 [f(x)] \cos [f(x)] f'(x)$

   Difficulty: 3   Section: 5

51. Making the usual assumptions, write a formula for the derivative of the function $f[\tan 2x]$.

   Answer: $2 \sec^2 (2x) f'[\tan 2x]$

   Difficulty: 3   Section: 5

52. Let $y = (x^2 + 1)^5$. Find $\dfrac{dy}{dx}$.

   Answer: $10x (x^2 + 1)^4$

   Difficulty: 1   Section: 5

53. Let $y = (x^3 - 2)^5$. Find $\dfrac{dy}{dx}$.

   Answer: $15x^2 (x^3 - 2)^4$

   Difficulty: 2   Section: 5

54. Let $y = (7 - x^3)^3$. Find $\dfrac{dy}{dx}$.

   Answer: $9x^2 (7 - x^3)^{-4}$

   Difficulty: 2   Section: 5

55. Let $y = \left(\dfrac{x^2 + 1}{x}\right)^4$. Compute $\dfrac{dy}{dx}$ and simplify.

   Answer: $\dfrac{4 (x^2 + 1)^3 (x^2 - 1)}{x^5}$

   Difficulty: 2   Section: 5

56. Let $y = \cos (x^3)$. Find $\dfrac{dy}{dx}$.

   Answer: $-3x^2 \sin (x^3)$

   Difficulty: 2   Section: 5

57. Let $y = \cos^3 (x)$. Find $\dfrac{dy}{dx}$.

Answer: $-3\cos^2(x)\sin(x)$

Difficulty: 2   Section: 5

58. Let $y = \sec^2(5x)$. Find $\dfrac{dy}{dx}$.

    Answer: $10\sec^2(5x)\tan(5x)$

    Difficulty: 2   Section: 5

59. Let $p = \tan(u^2)$. Find $\dfrac{dp}{du}$.

    Answer: $2u\sec^2(u^2)$

    Difficulty: 2   Section: 5

60. Let $y = x^2\cos 2x$. Find $\dfrac{dy}{dx}$.

    Answer: $2x(\cos 2x - x\sin 2x)$

    Difficulty: 2   Section: 5

61. The radius of a circle at time t is given by $r(t) = (t^2 + t)^2$. Find the rate of change of the area of the circle with respect to time.

    Answer: $4\pi(t^2 + t)^3(2t + 1)$

    Difficulty: 2   Section: 5

62. Let $f(x) = x\sqrt{1 + x^3}$. Find the value of $f'(x)$ when $x = 2$.

    Answer: 7

    Difficulty: 2   Section: 5

63. Find the equation of the tangent line to the curve $y = \sqrt{25 - x^2}$ at the point $(3, 4)$.

    Answer: $3x + 4y = 25$

    Difficulty: 2   Section: 5

64. Let $f(x) = \sqrt[3]{5x - 8}$. Where is the equation of the tangent line to the graph of f at the point where $x = 7$?

    Answer: $5x - 27y + 46 = 0$

    Difficulty: 2   Section: 5

65. Let $f(x) = \dfrac{1 + x}{x}$. Find $f'(x)$ and $f''(x)$.

    Answer: $-\dfrac{1}{x^2}$; $\dfrac{2}{x^3}$

Difficulty: 1   Section: 6

66. Let $y = x^6 - 3x^4 + 2x^2 - 3x + 10$. Find $\dfrac{d^2y}{dx^2}$.

   Answer: $30x^4 - 36x^2 + 4$

   Difficulty: 1   Section: 6

67. Let $p = q^{-2} + q^{-1}$. Find $\dfrac{d^3p}{d^3q}$.

   Answer: $-24q^{-5} - 6q^{-4}$

   Difficulty: 2   Section: 6

68. If $f(x) = x^7 + 3x^5 + x$, find the value of $f^{(8)}(x)$.

   Answer: 0

   Difficulty: 2   Section: 6

69. At time t, the position of a body moving in a line is given by $s = 3 + 5t - t^2$. What is the velocity at $t = 3$?

   Answer: -1

   Difficulty: 1   Section: 6

70. At time t, the position of a body moving in a line is given by $s = 3t^8 - 2t^5$. Find the velocity and acceleration as functions of time.

   Answer: $v(t) = 24t^7 - 10t^4$, $a(t) = 168t^6 - 40t^3$

   Difficulty: 1   Section: 6

71. Let $y = 2\sin(x) - \dfrac{1}{x^2}$. Find $\dfrac{d^2y}{dx^2}$.

   Answer: $-2\sin(x) - 3x^{-4}$.

   Difficulty: 2   Section: 6

72. Find $f''(x)$ if $f(x) = \sec(x)$.

   Answer: $\sec^3(x) + \sec(x)\tan^2(x)$

   Difficulty: 2   Section: 6

73. The function $s = 3 + t^2 - 2\cos(t)$ specifies the position of a particle at time t. Find the acceleration at time $t = \dfrac{\pi}{6}$.

   Answer: $2 + \sqrt{3}$

   Difficulty: 2   Section: 6

74. Let $y = (1-x)^4$. Find $\dfrac{d^2y}{dx^2}$.

   Answer: $12(1-x)^2$

   Difficulty: 1   Section: 6

75. Let $y = \dfrac{x}{1+x}$. Find $\dfrac{d^2y}{dx^2}$.

   Answer: $\dfrac{-2}{(1+x)^3}$

   Difficulty: 2   Section: 6

76. Let $g(x) = (x-1)^5 + (x+1)^2 + 2$. Find $g^{(4)}(x)$.

   Answer: 240

   Difficulty: 2   Section: 6

77. Let $f(x) = \dfrac{3}{20}(2-5x)^{-4/3}$. Find $f'(x)$ and $f''(x)$.

   Answer: $f'(x) = (2-5x)^{-7/3}$ and $f''(x) = \dfrac{35}{3}(2-5x)^{-10/3}$

   Difficulty: 2   Section: 6

78. The function $s = \sin(2x^2) - \pi$ gives the position at time t of a particle. Find the acceleration at any time t.

   Answer: $4\cos(2x^2) - 16x^2\sin(2x^2)$

   Difficulty: 2   Section: 6

79. Use implicit differentiation to find $\dfrac{dy}{dx}$ if $x^2y + xy^2 = 5y$.

   Answer: $\dfrac{2xy + y^2}{5 - x^2 - 2xy}$

   Difficulty: 2   Section: 7

80. Use implicit differentiation to find $\dfrac{dy}{dx}$ if $x^2 - xy + y^2 = 3$.

   Answer: $\dfrac{y - 2x}{2y - x}$

   Difficulty: 2   Section: 7

81. Use implicit differentiation to find $\dfrac{dy}{dx}$ if $x^2 + xy + y^5 = 2$.

   Answer: $-\dfrac{2x + y}{x + 5y^4}$

Difficulty: 2   Section: 7

82. What is the slope of the tangent line to the curve $4x^2 - 9y^2 = 36$ at the point $(6, 2\sqrt{3})$?

    Answer: $\dfrac{4\sqrt{3}}{9}$

    Difficulty: 2   Section: 7

83. Find the slope of the tangent line to the curve $x^2 - xy + y^2 = 3$ at the point $(-1, 1)$.

    Answer: 1

    Difficulty: 2   Section: 7

84. Find the equation of the tangent line to the curve $x^2 - xy + y^2 = 3$ at the point $(-1, 1)$?

    Answer: $x - y + 2 = 0$

    Difficulty: 2   Section: 7

85. Find the slope of the tangent line to the curve $xy + y^3 = 12$ at the point $(2, 2)$?

    Answer: $\dfrac{1}{14}$

    Difficulty: 2   Section: 7

86. Find an equation for the tangent line to the curve $x^3 - x + xy + y^3 + 24 = 0$ at the point $(-1, -3)$.

    Answer: $y + 3 = \dfrac{1}{26}(x + 1)$ or $x - 26y - 77 = 0$

    Difficulty: 2   Section: 7

87. Use implicit differentiation to find $\dfrac{du}{dw}$ if $w^3 - 2 + wu + u^3 + 24 = 0$.

    Answer: $\dfrac{1 - 3w^2 - u}{w + 3u^2}$

    Difficulty: 2   Section: 7

88. Use implicit differentiation to find $\dfrac{ds}{dt}$ if $s^2 - ts = 10$.

    Answer: $\dfrac{s}{2s - t}$

    Difficulty: 2   Section: 7

89. Use implicit differentiation to find $\dfrac{dr}{dw}$ if $5w^3 + 2wr^2 + r^3 = 10$.

    Answer: $-\dfrac{15w^2 + 2r^2}{4wr + 3r^2}$

Difficulty: 2   Section: 7

90. Use implicit differentiation to find $\dfrac{du}{dw}$ if $w^2 - wu + u^2 = 3$.

   Answer: $\dfrac{u - 2w}{2u - w}$

   Difficulty: 2   Section: 7

91. Use implicit differentiation to find $\dfrac{ds}{dt}$ if $5t^2 + 2t^2 s + s^2 = 1$.

   Answer: $-\dfrac{5t + 2ts}{t^2 + s}$

   Difficulty: 2   Section: 7

92. Use implicit differentiation to find $\dfrac{dy}{dx}$ if $y = x \cos y$.

   Answer: $\dfrac{\cos y}{1 + x \sin y}$

   Difficulty: 2   Section: 7

93. Use implicit differentiation to find $\dfrac{dy}{dx}$ if $y = x \sin y$.

   Answer: $\dfrac{\sin y}{1 - x \cos y}$

   Difficulty: 2   Section: 7

94. Use implicit differentiation to find $\dfrac{dy}{dx}$ if $x + \cos y - y^2 = 17$.

   Answer: $\dfrac{1}{2y + \sin y}$

   Difficulty: 2   Section: 7

95. Use implicit differentiation to find $\dfrac{dy}{dx}$ if $\tan y - y^3 + x^2 + 20 = 0$.

   Answer: $\dfrac{2x}{3y^2 - \sec^2 y}$

   Difficulty: 2   Section: 7

96. Use implicit differentiation to find $\dfrac{dy}{dx}$ if $y^2 - x^2 - \cos(xy) = 0$.

   Answer: $\dfrac{2x - y \sin(xy)}{2y + x \sin(xy)}$

   Difficulty: 2   Section: 7

97. Use implicit differentiation to find $\dfrac{dy}{dx}$ if $\sin^2(2y) - y^2 = 3x^2$.

   Answer: $\dfrac{3x^2}{4\sin(2y) - 2y}$

   Difficulty: 2   Section: 7

98. Use implicit differentiation to find $\dfrac{dy}{dx}$ if $\cos(yx) = x$.

   Answer: $\dfrac{y\sin(yx) + 1}{x\sin(yx)}$

   Difficulty: 2   Section: 7

99. Use implicit differentiation to find $\dfrac{dy}{dx}$ if $\sin x + \sin y = xy$.

   Answer: $\dfrac{y - \cos x}{\cos y - x}$

   Difficulty: 2   Section: 7

100. Use implicit differentiation to find $\dfrac{dy}{dx}$ if $\cos 3y = \tan 2x$.

   Answer: $-\dfrac{2\sec^2(2x)}{3\sin 3y}$

   Difficulty: 2   Section: 7

101. Use implicit differentiation to find $\dfrac{dy}{dx}$ and $\dfrac{d^2y}{dx^2}$ if $x^2 y^3 = 1$.

   Answer: $\dfrac{dy}{dx} = -\dfrac{2y}{3x},\ \dfrac{d^2y}{dx^2} = \dfrac{10y}{9x^2}$.

   Difficulty: 3   Section: 7

102. Use implicit differentiation to find $\dfrac{dy}{dx}$ and $\dfrac{d^2y}{dx^2}$ if $y^3 = x^2 - 17$.

   Answer: $\dfrac{dy}{dx} = \dfrac{2x}{3y^2}$ and $\dfrac{d^2y}{dx^2} = \dfrac{6y^3 - 8x^2}{9y^5}$

   Difficulty: 3   Section: 7

103. Use implicit differentiation to find $\dfrac{dy}{dx}$ if $x^{2/3} + y^{2/3} = 4$.

   Answer: $-\left(\dfrac{y}{x}\right)^{1/3}$

   Difficulty: 2   Section: 7

104. Use implicit differentiation to find $\dfrac{dy}{dx}$ if $3x^{4/3} + xy + 3y^{4/3} = 2$.

Answer: $\dfrac{4x^{1/3}+y}{x+4y^{1/3}}$

Difficulty: 2   Section: 7

105. Use implicit differentiation to find $\dfrac{dy}{dx}$ if $\sqrt{x+y}=x$.

Answer: $2\sqrt{x+y}-1$

Difficulty: 2   Section: 7

106. Use implicit differentiation to find $\dfrac{dy}{dx}$ if $2(xy)^{3/2}+y^2=x^2=0$.

Answer: $\dfrac{2x-3y\sqrt{xy}}{3x\sqrt{xy}+2y}$

Difficulty: 3   Section: 7

107. The radius of a circle is increasing at the rate of 3 ft/sec. Find the rate of change of the area of the circle at the time that the radius is 15 ft.

Answer: 90π ft^2/sec.

Difficulty: 1   Section: 8

108. A spherical balloon is inflated at a rate of 4 ft^3/min. How fast is the radius increasing when the radius is 2 ft?

Answer: $\dfrac{1}{4\pi}$ ft/min

Difficulty: 1   Section: 8

109. A ladder 20 feet long leans against a vertical building. If the bottom of the ladder slides away from the building horizontally at a rate of 2 ft/sec, how fast is the ladder sliding down the building when the top of the ladder is 12 feet above the ground?

Answer: $\dfrac{8}{3}$ ft/sec

Difficulty: 2   Section: 8

110. A 16ft ladder is leaning against a wall. Suppose the bottom of the ladder is pulled along the level sidewalk, directly away from the wall, at 2 ft/sec. How fast is the height of the midpoint of the ladder decreasing when the ladder is 4 ft from the wall?

Answer: $\dfrac{\sqrt{15}}{15}$ ft/sec

Difficulty: 2   Section: 8

111. A two-piece extension ladder leaning against a wall is collapsing at a rate of 2 ft/sec while the foot of the ladder remains a constant 5 ft from the wall. How fast is the ladder moving down the wall when the ladder is 13 ft long?

Answer: $\dfrac{13}{6}$ ft/sec

Difficulty: 2   Section: 8

112. A balloon is rising vertically at a rate of 2 ft/sec. An observer is located 300 ft from a point on the ground directly below the balloon. At what rate is the distance between the balloon and the observer changing when the height of the balloon is 400 ft?

Answer: 1.6 ft/sec

Difficulty: 2   Section: 8

113. A car leaves Albuquerque and travels east at 36 mph. One hour later, a second car traveling southwards at a constant speed is 48 miles south of Albuquerque and the distance between them is increasing at a rate of $\dfrac{592}{15}$ mph. Determine the speed of the second car.

Answer: $\dfrac{67}{3}$ mph

Difficulty: 2   Section: 8

114. A conical tank has radius 3 ft and depth 10 ft. If water is poured into the tank at the rate of 2 ft^3/min, how fast is the water level rising when the water in the tank is 5 ft deep?

Answer: $\dfrac{8}{9\pi}$ ft/min

Difficulty: 2   Section: 8

115. Sand is pouring from a pipe at the rate of 12ft^3/sec. If the falling sand forms a conical pile on the ground whose height is always $\dfrac{1}{3}$ the diameter of the base, how fast is the height increasing when the pile is 4 ft high? The volume of a cone is $V = \dfrac{1}{3}\pi r^2 h$ where r is the radius of the base and h is the height.

Answer: $\dfrac{1}{3\pi}$ ft/sec

Difficulty: 2   Section: 8

116. A baseball player runs from home plate to first base at a rate of 25 ft/sec. At what rate is his distance from third base changing when he is 40 ft from first base? A baseball diamond is 90 ft by 90 ft square.

Answer: $\dfrac{125}{\sqrt{106}}$ ft/sec

Difficulty: 2   Section: 8

117. The top of a tower has the shape of a hemisphere of radius 5 ft. The top has a uniform coating of ice which is 3 inches thick. The thickness of the ice is decreasing at the rate of $\dfrac{1}{3}$ in/hr. How fast is the volume of the ice changing? The volume of a sphere is $\dfrac{4}{3}\pi r^3$ where r

is the radius of the sphere.

Answer: 2646π in^3/hr

Difficulty: 2   Section: 8

118. A trough 10 ft long has a cross section that is an isosceles triangle having a base of 4 ft and height 3 ft. If the water pours into the trough at 6 ft^3/min, how fast is the depth of the water changing when the depth is 1 ft?

Answer: 0.45 ft/min

Difficulty: 2   Section: 8

119. A beacon 0.5 miles (perpendicular distance) from a straight shore revolves at 2 revolutions per minute (4π radians per minute). At what speed is the beam moving along the shore when it hits the shore 1 mile from the lighthouse?

Answer: 8π miles/minute

Difficulty: 2   Section: 8

120. In a football game, a receiver runs north at a rate of 30 ft/sec while the quarterback runs east at a rate of 20 ft/sec. How fast are the two players moving away from each other when the receiver is 80 ft down field and the quarterback is 60 ft east of center?

Answer: 36 ft/sec

Difficulty: 2   Section: 8

121. Let $f(x) = \dfrac{1}{x^2}$. Find a general expression for dy. Evaluate dy if $x = -2$ and $dx = 0.5$.

Answer: $dy = -\dfrac{2}{x^3}$, $dx = 0.125$

Difficulty: 1   Section: 9

122. Let $f(x) = x\cos(x)$. Find a general expression for dy. Evaluate dy if $x = \pi$ and $dx = 0.15$.

Answer: $[\cos(x) - x\sin(x)]\, dx, 0.15$

Difficulty: 1   Section: 9

123. Let $f(x) = \sec(x)$. Find a general expression for dy. Evaluate dy if $x = \dfrac{\pi}{3}$ and $dx = -0.5$.

Answer: $[\sec(x)\tan(x)]\, dx, -\sqrt{3}$

Difficulty: 1   Section: 9

124. Let $f(x) = \sin(x) - 4x$. Find a general expression for dy. Evaluate dy if $x = 0$, and $dx = 0.3$.

Answer: $[\cos(x) - 4]\, dx, -0.9$

Difficulty: 1   Section: 9

125. Let $y = 3x^2 - 5$. Compute dy and Δy if x changes from 2 to 2.1.

   Answer: $dy = 1.2$, $\Delta y = 1.23$

   Difficulty: 2   Section: 9

126. Let $y = \dfrac{1}{2 - x^3}$. Use dy to approximate the change in y if x changes from 1 to 1.02.

   Answer: $\Delta y \approx dy = 0.06$

   Difficulty: 1   Section: 9

127. Approximate $\sqrt{98}$ by means of the differential.

   Answer: 9.9

   Difficulty: 2   Section: 9

128. Approximate $\sqrt[3]{26}$ by means of the differential. Round off answer to three decimal places.

   Answer: 2.963

   Difficulty: 2   Section: 9

129. The radius of a circle is measured as 17 cm with an error of ± 0.4 cm. Estimate the error in calculating the circumference of the circle

   Answer: $\pm 0.8\pi$ cm

   Difficulty: 2   Section: 9

130. The side of a cube is measured at 10 cm with an error of ± 0.2 cm. Estimate the error in calculating the surface area of the cube.

   Answer: ± 24 square cm

   Difficulty: 2   Section: 9

131. Estimate the error in computing $\left(x^3 - 2x\right)^3$ at $x = 2$ if the error in x is estimated at ± 0.02.

   Answer: ± 9.6

   Difficulty: 2   Section: 9

132. Estimate the error in computing $2 + \sin^3 x$ at $x = \dfrac{\pi}{4}$ if the error in x is estimated at ± 0.2.

   Answer: $\pm (0.15)\sqrt{2}$

   Difficulty: 2   Section: 9

3 Applications of the Derivative

1. Determine the maximum and the minimum of the function $f(x) = x^3 - 3x^2 + 2$ on the interval $[-1, 1]$.

   Answer:
   maximum: 2
   minimum: -2

   Difficulty: 1   Section: 1

2. Determine the maximum and the minimum of the function $f(x) = x^3 - 6x^2 + 9x + 3$ on the interval $[1, 5]$.

   Answer:
   maximum: 23
   minimum: 3

   Difficulty: 1   Section: 1

3. Determine the maximum and the minimum of the function $f(x) = \dfrac{8x}{x^2 + 4}$ on the interval $[0, 3]$.

   Answer:
   maximum: 2
   minimum: 0

   Difficulty: 2   Section: 1

4. Determine the maximum and the minimum of the function $f(x) = 4x^3 - x^4$ on the interval $[-1, 2]$.

   Answer:
   maximum: 16
   minimum: -5

   Difficulty: 2   Section: 1

5. Determine the maximum and the minimum of the function $f(x) = x^4 - 4x^3 + 10$ on the interval $[0, 4]$.

   Answer:
   maximum: 10
   minimum: -17

   Difficulty: 2   Section: 1

6. Determine the maximum and the minimum of the function $f(x) = x^3 - 12x$ on the interval $[0, 3]$.

   Answer:
   maximum: 0

minimum: -16

Difficulty: 2   Section: 1

7. Determine the maximum and the minimum of the function $f(x) = x^3 - 3x + 2$ on the interval $[0, 2]$.

   Answer:
   maximum: 4
   minimum: 0

   Difficulty: 1   Section: 1

8. Determine the maximum and the minimum of the function $f(x) = \dfrac{x}{x^2 + 1}$ on the interval $[-1, 0]$.

   Answer:
   maximum: 0
   minimum: $-\dfrac{1}{2}$

   Difficulty: 2   Section: 1

9. Determine the maximum and the minimum of the function $f(x) = \dfrac{1}{3}x^3 + \dfrac{1}{2}x^2 - 6x + 8$ on the interval $[-4, 0]$.

   Answer:
   maximum: $\dfrac{43}{2}$
   minimum: 8

   Difficulty: 2   Section: 1

10. Determine the maximum and the minimum of the function $f(x) = x^{2/3}$ on the interval $[-8, 1]$.

    Answer:
    maximum: 4
    minimum: 0

    Difficulty: 1   Section: 1

11. Determine the maximum and the minimum of the function $f(x) = -\dfrac{4}{x^2}$ on the interval $[-3, -1]$.

    Answer:
    maximum: -3
    minimum: -5

    Difficulty: 2   Section: 1

12. Determine the maximum and the minimum of the function $f(x) = x(1-x)^{2/5}$ on the interval $\left[\dfrac{1}{2}, 2\right]$.

Answer:
>   maximum: 2
>   minimum: 0

Difficulty: 2   Section: 1

13. Determine the maximum and the minimum of the function $f(x) = x^3 - 3x^2$ on the interval $[1, 2]$.

    Answer:
    >   maximum: -2
    >   minimum: -4

    Difficulty: 2   Section: 1

14. Determine the intervals where $f(x) = \frac{1}{3}x^3 + x^2 - 3x$ is increasing.

    Answer: $(-\infty, -3], [1, \infty)$

    Difficulty: 2   Section: 2

15. Determine the intervals where $f(x) = \frac{1}{3}x^3 + x^2 - 3x$ is decreasing.

    Answer: $[-3, 1]$

    Difficulty: 2   Section: 2

16. Determine the intervals where $f(x) = \frac{1}{1+x^2}$ is increasing.

    Answer: $(-\infty, 0]$

    Difficulty: 2   Section: 2

17. Determine the intervals where $f(x) = x^4 - 4x^3 + 10$ is increasing.

    Answer: $[3, 0]$

    Difficulty: 2   Section: 2

18. Determine the intervals where $f(x) = \frac{x}{x^2+1}$ is increasing.

    Answer: $[-1, 1]$

    Difficulty: 2   Section: 2

19. Determine the intervals where $f(x) = \frac{1}{6}\left(x^3 - 6x^2 + 9x + 6\right)$ is increasing.

    Answer: $(-\infty, 1], [3, \infty)$

    Difficulty: 2   Section: 2

20. Determine the intervals where $f(x) = \frac{8x}{x^2+4}$ is decreasing.

Answer: $(-\infty, -2], [2, \infty)$

Difficulty: 2   Section: 2

21. Determine the intervals where $f(x) = x^3 - 3x^2 + 2$ is decreasing.

    Answer: $[0, 2]$

    Difficulty: 1   Section: 2

22. Determine the intervals where $f(x) = 4x^3 - x^4$ is increasing.

    Answer: $(-\infty, 3]$

    Difficulty: 2   Section: 2

23. Determine the intervals where $f(x) = x^3 - 6x^2 + 9x + 3$ is increasing.

    Answer: $(-\infty, 1], [3, \infty)$

    Difficulty: 2   Section: 2

24. Determine the intervals where $f(x) = x^2 - 4x + 5$ is decreasing.

    Answer: $(-\infty, 2]$

    Difficulty: 1   Section: 2

25. Determine the intervals where $f(x) = \frac{1}{3}x^3 - \frac{1}{2}x^2 - 6x - 4$ is decreasing.

    Answer: $[-2, 3]$

    Difficulty: 2   Section: 2

26. Determine the intervals where $f(x) = \frac{1}{3}x^3 + \frac{1}{2}x^2 - 6x + 8$ is decreasing.

    Answer: $[-3, 2]$

    Difficulty: 2   Section: 2

27. Determine the intervals where $f(x) = x^2 - 7x + 12$ is increasing.

    Answer: $[\frac{7}{2}, \infty)$

    Difficulty: 1   Section: 2

28. Let $f(x) = x^3 - 3x^2 + 2$. Determine the intervals where $f(x)$ is concave up, intervals where $f(x)$ is concave down, and find all points of inflection.

    Answer:
    concave up: $(1, \infty)$
    concave down: $(-\infty, 1)$
    points of inflection: $(1, 0)$

Difficulty: 1   Section: 2

29. Let $f(x) = x^3 - 12x$. Determine the intervals where $f(x)$ is concave up, intervals where $f(x)$ is concave down, and find all points of inflection.

   Answer:
   concave up: $(0, \infty)$
   concave down: $(-\infty, 0)$
   points of inflection: $(0, 0)$

   Difficulty: 1   Section: 2

30. Let $f(x) = x - 3 + \dfrac{2}{x+1}$. Determine the intervals where $f(x)$ is concave up, intervals where $f(x)$ is concave down, and find all points of inflection.

   Answer:
   concave up: $(-1, \infty)$
   concave down: $(-\infty, -1)$
   points of inflection: none

   Difficulty: 2   Section: 2

31. Let $f(x) = \dfrac{x}{x^2+1}$. Determine the intervals where $f(x)$ is concave up, intervals where $f(x)$ is concave down, and find all points of inflection.

   Answer:
   concave up: $(-\sqrt{3}, 0), (\sqrt{3}, \infty)$
   concave down: $(-\infty, -\sqrt{3}), (0, \sqrt{3})$
   points of inflection: $(-\sqrt{3}, -\sqrt{3}/4), (0, 0), (\sqrt{3}, \sqrt{3}/4)$

   Difficulty: 2   Section: 2

32. Let $f(x) = 4x^3 - x^4$. Determine the intervals where $f(x)$ is concave up, intervals where $f(x)$ is concave down, and find all points of inflection.

   Answer:
   concave up: $(0, 2)$
   concave down: $(-\infty, 0), (2, \infty)$
   points of inflection: $(0, 0), (2, 16)$

   Difficulty: 1   Section: 2

33. Let $f(x) = \dfrac{x^3}{(x-2)^3}$. Determine the intervals where $f(x)$ is concave up, intervals where $f(x)$ is concave down, and find all points of inflection.

   Answer:
   concave up: $(-2, 0), (2, \infty)$
   concave down: $(-\infty, -2), (0, 2)$
   points of inflection: $(0, 0), (-2, 1/8)$

Difficulty: 2   Section: 2

34. Let $f(x) = \frac{1}{3}x^3 - \frac{1}{2}x^2 - 6x - 4$. Determine the intervals where $f(x)$ is concave up, intervals where $f(x)$ is concave down, and find all points of inflection.

    Answer:
    concave up: $\left(\frac{1}{2}, \infty\right)$
    concave down: $\left(\infty, \frac{1}{2}\right)$
    points of inflection: $\left(\frac{1}{2}, -\frac{85}{12}\right)$

    Difficulty: 2   Section: 2

35. Let $f(x) = \frac{1}{6}\left(x^3 - 6x^2 + 9x + 6\right)$. Determine the intervals where $f(x)$ is concave up, intervals where $f(x)$ is concave down, and find all points of inflection.

    Answer:
    concave up: $(2, \infty)$
    concave down: $(-\infty, 2)$
    points of inflection: $(2, 5/6)$

    Difficulty: 2   Section: 2

36. Let $f(x) = x^3 - 6x^2 + 9x + 3$. Determine the intervals where $f(x)$ is increasing, intervals where $f(x)$ is decreasing, intervals where $f(x)$ is concave up, intervals where $f(x)$ is concave down, and find all points of inflection.

    Answer:
    increasing: $(\infty, 1], [3, \infty)$
    decreasing: $[1, 3]$
    concave up: $(2, \infty)$
    concave down: $(-\infty, 2)$
    points of inflection: $(2, 5)$

    Difficulty: 2   Section: 2

37. Let $f(x) = x^3 - 3x^2 + 1$. Determine the intervals where $f(x)$ is increasing, intervals where $f(x)$ is decreasing, intervals where $f(x)$ is concave up, intervals where $f(x)$ is concave down, and find all points of inflection.

    Answer:
    increasing: $(-\infty, 0], [2, \infty)$
    decreasing: $[0, 2]$
    concave up: $(1, \infty)$
    concave down: $(-\infty, 1)$
    points of inflection: $(1, -1)$

    Difficulty: 2   Section: 2

38. Let $f(x) = x^4 - 4x^3 + 10$. Determine the intervals where $f(x)$ is increasing, intervals where

$f(x)$ is decreasing, intervals where $f(x)$ is concave up, intervals where $f(x)$ is concave down, and find all points of inflection.

Answer:
increasing: $[3, \infty)$
decreasing: $(-\infty, 3]$
concave up: $(-\infty, 0), (2, \infty)$
concave down: $(0, 2)$
points of inflection: $(0, 10), (2, -6)$

Difficulty: 2   Section: 2

39. Let $f(x) = \dfrac{8x}{x^2+4}$. Determine the intervals where $f(x)$ is increasing, intervals where $f(x)$ is decreasing, intervals where $f(x)$ is concave up, intervals where $f(x)$ is concave down, and find all points of inflection.

Answer:
increasing: $[-2, 2]$
decreasing: $(-\infty, -2], [2, \infty)$
concave up: $(-2\sqrt{3}, 0), (2\sqrt{3}, \infty)$
concave down: $(-\infty, -2\sqrt{3}), (0, 2\sqrt{3})$
points of inflection: $(-2\sqrt{3}, -\sqrt{3}), (0, 0), (2\sqrt{3}, \sqrt{3})$

Difficulty: 2   Section: 2

40. Find all points which give a local maximum and all points which give a local minimum value for the function $f(x) = x^3 - 6x^2 + 9x + 3$.

Answer:
local maximum: at $x = 1$
local minimum: at $x = 3$

Difficulty: 2   Section: 3

41. Find all points which give a local maximum and all points which give a local minimum value for the function $f(x) = x^3 - 3x^2 + 2$.

Answer:
local maximum: at $x = 0$
local minimum: at $x = 2$

Difficulty: 1   Section: 3

42. Find all points which give a local maximum and all points which give a local minimum value for the function $f(x) = x^4 - 4x^3 + 10$.

Answer:
local maximum: none
local minimum: at $x = 3$

Difficulty: 2   Section: 3

43. Find all points which give a local maximum and all points which give a local minimum value for the function $f(x) = x^3 - 12x$.

   Answer:
   local maximum: at $x = -2$
   local minimum: at $x = 2$

   Difficulty: 2   Section: 3

44. Find all points which give a local maximum and all points which give a local minimum value for the function $f(x) = x - 3 + \dfrac{2}{x+1}$

   Answer:
   local maximum: at $x = -1 - \sqrt{2}$
   local minimum: at $x = -1 + \sqrt{2}$

   Difficulty: 2   Section: 3

45. Find all points which give a local maximum and all points which give a local minimum value for the function $f(x) = \dfrac{8x}{x^2 + 4}$

   Answer:
   local maximum: at $x = 2$
   local minimum: at $x = -2$

   Difficulty: 2   Section: 3

46. Find all points which give a local maximum and all points which give a local minimum value for the function $f(x) = x^3 - 3x^2 + 1$.

   Answer:
   local maximum: at $x = 0$
   local minimum: at $x = 2$.

   Difficulty: 1   Section: 3

47. Find all points which give a local maximum and all points which give a local minimum value for the function $f(x) = \dfrac{1}{1 + x^2}$

   Answer:
   local maximum: at $x = 0$
   local minimum: none

   Difficulty: 2   Section: 3

48. Find all points which give a local maximum and all points which give a local minimum value for the function $f(x) = \dfrac{x}{1 + x^2}$.

   Answer:
   local maximum: at $x = 1$
   local minimum: at $x = -1$

Difficulty: 2   Section: 3

49. Find all points which give a local maximum and all points which give a local minimum value for the function $f(x) = \frac{1}{6}\left(x^3 - 6x^2 + 9x + 6\right)$.

   Answer:
   local maximum: at $x = 1$
   local minimum: at $x = 3$

   Difficulty: 2   Section: 3

50. Find all points which give a local maximum and all points which give a local minimum value for the function $f(x) = 4x^3 - x^4$.

   Answer:
   local maximum: at $x = 3$
   local minimum: none

   Difficulty: 2   Section: 3

51. Find all points which give a local maximum and all points which give a local minimum value for the function $f(x) = \dfrac{x^3}{(x-2)^2}$.

   Answer:
   local maximum: none
   local minimum: none

   Difficulty: 2   Section: 3

52. Find all points which give a local maximum and all points which give a local minimum value for the function $f(x) = \frac{1}{3}x^3 - \frac{1}{2}x^2 - 6x - 4$.

   Answer:
   local maximum: at $x = -2$
   local minimum: at $x = 3$

   Difficulty: 2   Section: 3

53. Find two positive numbers whose sum is 17 and whose product is a maximum.

   Answer: $\dfrac{17}{2}, \dfrac{17}{2}$

   Difficulty: 2   Section: 1

54. A rectangular field is to be enclosed by a fence and divided into 3 lots by fences parallel to one of its sides. Find the dimensions of the largest field that can be enclosed with a total of 800 meters of fencing.

   Answer: 100 m by 200 m

   Difficulty: 2   Section: 4

55. A farmer wishes to enclose a 4000 square meter field and subdivide the field into four rectangular plots with fences parallel to one of the sides. What should the dimensions of the field be in order to minimize the amount of fencing required?

    Answer: 100 m by 40 m

    Difficulty: 2   Section: 4

56. An open box is formed from a square sheet of cardboard by cutting equal squares from each corner and folding up the edges. If the dimensions of the cardboard are 18 cm by 18 cm, what should be the dimensions of the box so as to maximize the volume, and what is the maximum volume?

    Answer: 3 cm by 12 cm by 12 cm. The maximum volume is 432 cm^3.

    Difficulty: 2   Section: 4

57. An open box is formed from a rectangular sheet of cardboard by cutting equal squares from each corner and folding up the edges. If the dimensions of the cardboard are 15 in by 24 in, what size squares should be cut to obtain a box of maximum volume?

    Answer: 3 in by 3 in

    Difficulty: 2   Section: 4

58. A box is made with a square bottom and open top to hold a volume of 8 ft^3. The material for the sides costs $4 per square foot and the material for the bottom costs $1 per square foot. What dimensions will minimize the cost of the box?

    Answer: Bottom 4 ft by 4 ft and height 0.5 ft

    Difficulty: 2   Section: 4

59. A car rental agency has 24 identical cars. The owner of the agency finds that at the price of $10 per day, all the cars can be rented. However, for each $1 increase in rental price, one of the cars is not rented. How much should be charged to maximize the income of the agency?

    Answer: $17

    Difficulty: 2   Section: 4

60. Suppose a rectangle has its lower base on the x-axis and upper vertices on the graph of the function $y = 8 - x^2$. Find the area of the largest such rectangle.

    Answer: $\dfrac{64\sqrt{6}}{9}$

    Difficulty: 2   Section: 4

61. A right triangle of hypotenuse 10 units is rotated about one of its legs to generate a right circular cone. What is the greatest possible volume of the cone?

    Answer: $\dfrac{2000\pi}{9\sqrt{3}} \approx 403$ units^3

Difficulty: 2   Section: 4

62. Find the base radius r and the height h of the right circular cone of maximum volume which will fit inside a sphere of radius 3 units.

    Answer: $r = 2\sqrt{2},\ h = 4$

    Difficulty: 2   Section: 4

63. A motorist is in a desert in a jeep. The closest point on a straight road is town A and it is $4\sqrt{2}$ miles from the motorist. He wishes to reach town B, 10 miles from A on the road, in the shortest time. If he can drive 15 mi/hr on the desert and 45 mi/hr on the road, where should he intersect the road?

    Answer: 2 mi from A

    Difficulty: 2   Section: 4

64. The total cost of producing and selling x units of a certain commodity per month is $C(x) = 4x^2 + 100x + 4000$. If the production level is 2000 units per month, find the average cost, $\dfrac{C(x)}{x}$, of each unit and the marginal cost.

    Answer: $8102, $16100

    Difficulty: 1   Section: 4

65. The total cost of producing x units is given by $C(x) = 3000 - 20x + 0.03x^2$. If the total revenue is $R(x) = 50x - 0.01x^2$, find the marginal profit when $x = 100$ units.

    Answer: $62

    Difficulty: 1   Section: 4

66. A manufacturer has total cost $C(x) = 2000 + 196x + 3x^2$ and total revenue is $R(x) = 500x - x^2$ when x units are manufactured per week. What is the marginal profit when $x = 10$? How many units should be manufactured for maximum profit? What is the maximum profit?

    Answer: $224/unit, 38, $3776

    Difficulty: 2   Section: 4

67. If x is the number of items produced in one week, the selling price per item is $(100 - 0.02x)$. The total cost for x items per week is $(40x + 15000)$. How many items should be produced and sold per week for maximum profit? What is the maximum profit?

    Answer: 1500, $30000

    Difficulty: 2   Section: 4

68. A shopping club charges its members $200 per year. However, for each new member in excess of 60, the charge for every member is reduced by $2. What number of members leads to a maximum revenue? What is the maximum revenue?

Answer: 80, $12800

Difficulty: 2 Section: 4

69. Sketch the graph of the function $f(x) = x^3 - 3x^2 + 2$.

    Answer:
    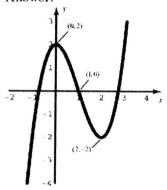

Difficulty: 1 Section: 5

70. Sketch the graph of the function $f(x) = x^3 - 6x^2 + 9x + 3$.

    Answer:
    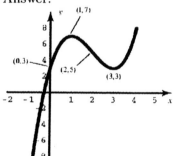

Difficulty: 1 Section: 5

71. Sketch the graph of the function $f(x) = \dfrac{8x}{x^2 + 4}$.

    Answer:
    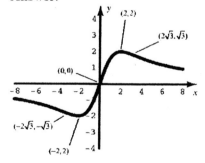

Difficulty: 2 Section: 5

72. Sketch the graph of the function $f(x) = \dfrac{x^3}{(x-2)^2}$.

    Answer:
    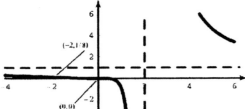

    Difficulty: 2   Section: 5

73. Sketch the graph of the function $f(x) = 4x^3 - x^4$.

    Answer:
    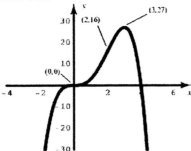

    Difficulty: 2   Section: 5

74. Sketch the graph of the function $f(x) = x^{2/3}$. Label the important functions.

    Answer:
    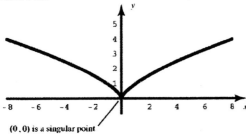

    Difficulty: 2   Section: 5

75. Sketch the graph of the function $f(x) = (x-2)^{1/3}$. Label the important features.

    Answer:
    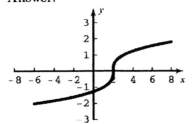

Difficulty: 2   Section: 5

76. Sketch the graph of the function $f(x) = x^2 - \dfrac{1}{|x|}$. Label the important features.

    Answer:

    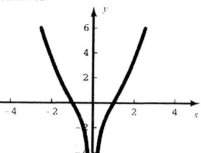

Difficulty: 2   Section: 5

77. Sketch the graph of the function $f(x) = \left(1 + x^5\right)^{-1}$. Label the important features.

    Answer:

    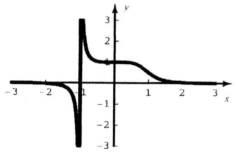

Difficulty: 3   Section: 5

78. Sketch the graph of the function $f(x) = x(1-x)^{2/5}$. Label the important features.

    Answer:

    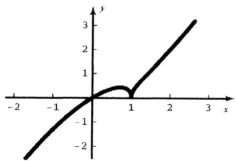

Difficulty: 3   Section: 5

79. Sketch the graph of the function $f(x) = 3x^{5/3} - 5x$. Label the important features.

    Answer:

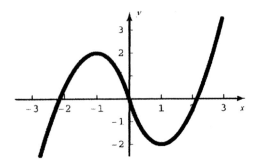

Difficulty: 2   Section: 5

Problems 80–84 refer to the graph of $y = f'(x)$ in the figure below.

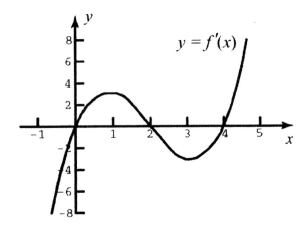

80. On what intervals is f increasing?

    Answer: $[0, 2]$, $[4, \infty)$

    Difficulty: 1   Section 5

81. On what intervals is f decreasing?

    Answer: $(-\infty, 0]$, $[2, 3]$

    Difficulty: 1   Section 5

82. On what intervals is f concave up?

    Answer: $(-\infty, 2)$, $(3, \infty)$

    Difficulty: 1   Section 5

83. Identify all local extrema of f?

    Answer: $f(0)$ and $f(4)$ are local minima; $f(2)$ is a local maximum.

    Difficulty: 1   Section 5

84. Sketch a graph of $y = f(x)$ assuming that $f(0) = 0$.

Answer:

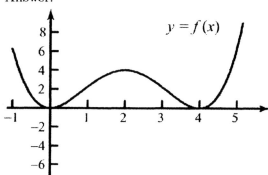

Difficulty: 1   Section 5

Problems 85–88 refer to the graph of $y = g'(x)$ in the figure below. The function g is continuous.

85. On what intervals is g increasing?

    Answer: $(-\infty, \infty)$

    Difficulty: 1   Section 5

86. On what intervals is g concave up?

    Answer: $(-\infty, 0)$

    Difficulty: 1   Section 5

87. On what intervals is g concave down?

    Answer: $(0, \infty)$

    Difficulty: 1   Section 5

88. Identify all local extrema of g.

    Answer: Since g is increasing on $(-\infty, \infty)$, g has no local extrema.

    Difficulty: 1   Section 5

89. Sketch a graph of $y = f(x)$ assuming that $f(0) = 0$.

    Answer:

    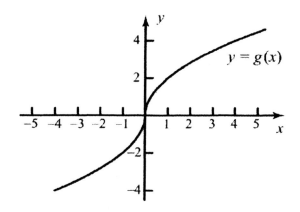

90. Let $f(x) = x^3$ on the interval $[0, 6]$. Find all numbers c, $0 < c < 6$, which satisfy the conclusion of the Mean Value Theorem.

    Answer: $c = 2\sqrt{3}$

    Difficulty: 1   Section: 6

91. Let $f(x) = \sqrt{2 - x^2}$ on the interval $[-\sqrt{2}, 0]$. Find all numbers c, $-\sqrt{2} < c < 0$, which satisfy the conclusion of the Mean Value Theorem.

    Answer: $c = -1$

    Difficulty: 2   Section: 6

92. Let $f(x) = \sqrt{2x - x^2}$ on the interval $[0, 2]$. Find all numbers c, $0 < c < 2$, which satisfy the conclusion of the Mean Value Theorem.

    Answer: $c = 1$

    Difficulty: 2   Section: 6

93. Use Newton's method to find an approximate solution of $x^3 + 3x - 20 = 0$ between 2 and 3. Round to two decimal places.

    Answer: 2.35

    Difficulty: 2   Section: 7

94. Use Newton's method to find an approximate solution of $3x^3 + 14x^2 + 13x - 2 = 0$ between 0 and 1. Round to two decimal places.

    Answer: 0.13

    Difficulty: 2   Section: 7

95. Use Newton's method to find an approximate solution of $x^4 - 2x - 100 = 0$ between 3 and 4. Round to two decimal places.

    Answer: 3.21

    Difficulty: 2   Section: 7

96. Use Newton's method to find a positive approximate solution of $x^3 - 144x + 665 = 0$. Round to two decimal places.

    Answer: There are two, 7 and 6.86.

    Difficulty: 2   Section: 7

97. In Newton's method to estimate the root of the equation $x^3 - x - 1 = 0$, take the first guess $x_0 = 1$. What is the second approximation x_1?

    Answer: 1.5

    Difficulty: 2   Section: 7

98. In Newton's method to estimate the root of the equation $x^2 - 5 = 0$, take the first guess $x_0 = 2$. What is the second approximation x_1?

   Answer: 2.25

   Difficulty: 2   Section: 7

99. Use the fixed-point method to solve the equation $x^3 + 3x - 20 = 0$ with initial guess of 2 and 4 iterations.

   Answer: the process diverges

   Difficulty: 2   Section: 7

100. Use the fixed-point method to solve the equation $3x^3 + 14x^2 + 13x - 2 = 0$ with initial guess of 0.5 and 4 iterations. Round off to two decimal places.

   Answer: 0.13

   Difficulty: 1   Section: 7

101. Use the fixed-point method to solve the equation $x^3 - 144x + 665 = 0$ with initial guess of 6.5 and 4 iterations. Round off to two decimal places.

   Answer: 6.58

   Difficulty: 1   Section: 7

102. Use the fixed-point method to solve the equation $x^3 - x - 1 = 0$ with initial guess of 2 and 4 iterations. Round off to two decimal places

   Answer: the process diverges

   Difficulty: 2   Section: 7

103. Find $\int 2x^2 \, dx$.

   Answer: $\dfrac{2}{3}x^3 + C$

   Difficulty: 1   Section: 8

104. Find $\int \left(-3x^{-4}\right) dx$.

   Answer: $x^{-3} + C$

   Difficulty: 1   Section: 8

105. Find $\int \dfrac{x^2 + 2x}{x} \, dx$

   Answer: $\dfrac{1}{2}x^2 + 2x + C$

   Difficulty: 1   Section: 8

106. Find $\int \dfrac{1}{\sqrt{x}}\, dx$.

   Answer: $2\sqrt{x} + C$

   Difficulty: 1   Section: 8

107. Find $\int \left(5x^4 + 3\right) dx$.

   Answer: $x^5 + 3x + C$

   Difficulty: 1   Section: 8

108. Find $\int \left(x^2 - \dfrac{1}{x^2}\right) dx$.

   Answer: $\dfrac{1}{3}x^3 + \dfrac{1}{x} + C$

   Difficulty: 1   Section: 8

109. Find $\int \left(7x^4 - x^{1/2} - 3\right) dx$.

   Answer: $\dfrac{7}{5}x^5 - \dfrac{2}{3}x^{3/2} - 3x + C$

   Difficulty: 1   Section: 8

110. Find $\int (9 - y)\, dy$.

   Answer: $9y - \dfrac{1}{2}y^2 + C$

   Difficulty: 1   Section: 8

111. Find $\int \left(x - \dfrac{1}{x}\right)^2 dx$.

   Answer: $\dfrac{1}{3}x^3 - 2x - x^{-1} + C$

   Difficulty: 2   Section: 8

112. Find $\int \left(t^2 - 1\right)^2 dt$.

   Answer: $\dfrac{1}{5}t^5 - \dfrac{2}{3}t^3 + t + C$

   Difficulty: 1   Section: 8

113. Find $\int (\sin u + \cos u - u)\, du$.

   Answer: $-\cos u + \sin u - \dfrac{1}{2}u^2 + C$

Difficulty: 2   Section: 8

114. Prove that $\int \sin^2 x \, dx = \dfrac{x}{2} - \dfrac{1}{4}\sin 2x + C$.

Answer:
Show that the derivative of the right side of the equation is equal to $\sin^2 x$. A trigonometric identity will be needed.

Difficulty: 3   Section: 8

115. Prove that $\int \cos^3 x \, dx = \sin x - \dfrac{\sin^3 x}{3} + C$.

Answer:
Show that the derivative of the right side of the equation is equal to $\cos^3 x$. A trigonometric identity will be needed.

Difficulty: 3   Section: 8

116. Prove that $\int u\left(2u^2 + 3\right)^3 du = \dfrac{1}{16}\left(2u^2 + 3\right)^4 + C$.

Answer:
Show that the derivative of the right side of the equation is $u\left(2u^2 + 3\right)^3$.

Difficulty: 2   Section: 8

117. Find $\int \left(x^3 + 2\right)^6 x^2 \, dx$.

Answer: $\dfrac{1}{21}\left(x^3 + 2\right)^7 + C$

Difficulty: 2   Section: 8

118. Find $\int \left(z^3 + 1\right)^2 dz$.

Answer: $\dfrac{1}{5}z^5 + \dfrac{2}{3}z^3 + z + C$

Difficulty: 2   Section: 8

119. Find $\int x^2 \left(2x^3 + 1\right)^7 dx$.

Answer: $\dfrac{1}{48}\left(2x^3 + 1\right)^8 + C$

Difficulty: 2   Section: 8

120. Find $\int 2x\left(4 - 7x^2\right)^{-7} dx$.

Answer: $\dfrac{1}{42}\left(4 - 7x^2\right)^{-6} + C$

Difficulty: 2   Section: 8

121. Find $\int \dfrac{1}{(3x+2)^2}\, dx$.

Answer: $-\dfrac{1}{3(3x+2)} + C$

Difficulty: 2   Section: 8

122. Find $\int \dfrac{1}{\sqrt{3x+2}}\, dx$.

Answer: $\dfrac{2}{3}\sqrt{3x+2} + C$

Difficulty: 2   Section: 8

123. Find $\int \dfrac{4y}{\sqrt{25-4y^2}}\, dy$.

Answer: $\sqrt{25-4y^2} + C$

Difficulty: 2   Section: 8

124. Find $\int x\sqrt[4]{1-x^2}\, dx$.

Answer: $-\dfrac{2}{5}\left(1-x^2\right)^{5/4} + C$

Difficulty: 2   Section: 8

125. Find $\int \sqrt{5-x}\, dx$.

Answer: $-\dfrac{2}{3}(5-x)^{3/2} + C$

Difficulty: 2   Section: 8

126. Find $\int x\cos\left(2x^2 - 1\right)\, dx$.

Answer: $\dfrac{1}{4}\sin\left(2x^2 - 1\right) + C$

Difficulty: 2   Section: 8

127. Find $\int x^3 \sin\left(4x^4 - \pi\right)\, dx$.

Answer: $-\dfrac{1}{16}\cos\left(4x^4 - \pi\right) + C$

Difficulty: 2   Section: 8

128. Find the solution to the differential equation $\dfrac{dy}{dx} = 2x + 3x^2$ satisfying $y = 5$ when $x = 1$.

   Answer: $y = x^2 + x^3 + 3$

   Difficulty: 1   Section: 9

129. Find the solution to the differential equation $\dfrac{dy}{dx} = 5x^4 + 3x^2 - 1$ satisfying $y = -7$ when $x = 1$.

   Answer: $y = x^5 + x^3 - x - 8$.

   Difficulty: 1   Section: 9

130. Find the solution to the differential equation $\dfrac{dy}{dx} = (x-2)(x+3)$ satisfying $y = -5$ when $x = 0$.

   Answer: $y = \dfrac{1}{3}x^3 + \dfrac{1}{2}x^2 - 6x - 5$

   Difficulty: 2   Section: 9

131. Find the solution to the differential equation $\dfrac{dy}{dp} = \sqrt{p}$ satisfying $y = 0$ when $p = 4$.

   Answer: $y = \dfrac{2}{3}p^{3/2} - \dfrac{16}{3}$

   Difficulty: 2   Section: 9

132. Find the solution to the differential equation $\dfrac{dy}{dt} = ty^2$ satisfying $y = 7$ at $t = 1$.

   Answer: $y = -\dfrac{2}{t^2} + 9$

   Difficulty: 2   Section: 9

133. Find the solution to the differential equation $\dfrac{dy}{dx} = \dfrac{4x}{y}$ satisfying $y = 4$ at $x = 1$.

   Answer: $y = 2\sqrt{x^2 + 3}$

   Difficulty: 2   Section: 9

134. Find the solution to the differential equation $\dfrac{ds}{dt} = (t^2 - 7)s^2$ satisfying $s = \dfrac{1}{6}$ at $t = 2$.

   Answer: $s = \dfrac{1}{7t - t^3}$.

   Difficulty: 2   Section: 9

135. Find the solution to the differential equation $\dfrac{dy}{dx} = y^2$ satisfying $y = 1$ at $x = 0$.

   Answer: $y = -\dfrac{1}{x - 1}$

Difficulty: 2   Section: 9

136. Find $g(t)$ if $g'(t) = \cos t$ and $g\left(\dfrac{\pi}{6}\right) = \dfrac{11}{2}$.

Answer: $\sin t + 5$

Difficulty: 2   Section: 9

137. Find $f(x)$ if $f'(x) = 2x - \dfrac{1}{x^2}$ and $f(2) = 4$.

Answer: $x^2 + \dfrac{1}{x} - \dfrac{1}{2}$

Difficulty: 2   Section: 9

138. A particle is moving along a line with velocity $v(t) = 3t^2 + t$. Find the position function $s(t)$ if $s(0) = 6$.

Answer: $t^3 + \dfrac{1}{2}t^2 + 6$

Difficulty: 1   Section: 9

139. A particle is moving along a line with velocity $v(t) = \sqrt{t} - 5$. Find the position function $s(t)$ if $s(0) = -7$.

Answer: $\dfrac{2}{3}t^{3/2} - 5t - 7$

Difficulty: 1   Section: 9

140. A particle is moving along a line with acceleration $a(t) = 3t^2 - 5$ feet/sec^2. Find the velocity and position functions if $v(0) = -5$ and $s(0) = 0$.

Answer: $v(t) = t^3 - 5t - 5$ ft/sec and $s(t) = \dfrac{1}{4}t^4 - \dfrac{5}{2}t^2 - 5t$ ft.

Difficulty: 2   Section: 9

141. A particle is moving along a line with acceleration $a(t) = \dfrac{1}{t^3} - t$ feet/sec^2. Find the velocity and position functions if $v(1) = 0$ and $s(1) = 0$.

Answer: $v(t) = -\dfrac{1}{2t^2} - \dfrac{1}{2}t^2 + 1$ ft/sec and $s(t) = \dfrac{1}{2t} - \dfrac{1}{6}t^3 + t - \dfrac{4}{3}$ ft.

Difficulty: 2   Section: 9

142. A ball is thrown straight downward from a height of 30 meters, with an initial velocity of 3 meters per second. When will the ball strike the ground?

Answer: $\dfrac{-3 + \sqrt{597}}{9.8} \approx 2.187$ seconds after it is thrown

Difficulty: 2   Section: 9

143. A ball is thrown straight downward from a height of 40 meters, with an initial velocity of 2 meters per second. When will the ball strike the ground?

Answer: $\dfrac{-2 + \sqrt{788}}{9.8} \approx 2.66$ seconds after it is thrown

Difficulty: 2   Section: 9

144. A bullet is shot straight upward from a point 5 feet above the ground. The muzzle velocity is 300 feet per second. At what time will the bullet reach its maximum height?

Answer: $\dfrac{75}{8} = 9.375$ seconds after firing.

Difficulty: 2   Section: 9

145. A bullet is shot straight upward from a point 5 feet above the ground. The muzzle velocity is 300 feet per second. What is the maximum height of the bullet?

Answer: $\dfrac{5645}{4} = 1411.25$ feet above the ground

Difficulty: 2   Section: 9

146. A bullet is shot straight upward from a point 5 feet above the ground. The muzzle velocity is 300 feet per second. At what time does the bullet strike the ground?

Answer: $\dfrac{300 + \sqrt{90320}}{32} \approx 18.77$ seconds after firing.

Difficulty: 2   Section: 9

147. A bullet is shot straight upward from ground level with a muzzle velocity of 256 feet per second. At what time will the bullet strike the ground?

Answer: 16 seconds after firing

Difficulty: 2   Section: 9

4 The Definite Integral

1. Compute the sum $\sum_{k=1}^{6}(3k-2)$.

   Answer: 51

   Difficulty: 1   Section: 1

2. Compute the sum $\sum_{k=0}^{6}(3k-10)$.

   Answer: -7

   Difficulty: 1   Section: 1

3. Compute the sum $\sum_{t=1}^{5} 4^t$.

   Answer: 1364

   Difficulty: 1   Section: 1

4. Compute the sum $\sum_{t=2}^{7}(t^2-t)$.

   Answer: 112

   Difficulty: 1   Section: 1

5. Compute the sum $\sum_{j=0}^{6}(-1)^{j+1} j^2$.

   Answer: -21

   Difficulty: 1   Section: 1

6. Compute the sum $\sum_{i=0}^{2}\left(\dfrac{i+2}{i+1}\right)^2$.

   Answer: $\dfrac{289}{36}$.

   Difficulty: 1   Section: 1

7. Compute the sum $\sum_{t=0}^{4}\left(\dfrac{t-t^2}{1+t}\right)$.

   Answer: $-\dfrac{137}{30}$

   Difficulty: 1   Section: 1

8. Compute the sum $\sum_{q=1}^{5}\left(\dfrac{3-q}{q}\right)^2$.

Answer: $\dfrac{1789}{400}$

Difficulty: 1   Section: 1

9. Compute the sum $\sum\limits_{j=0}^{3} \left(\dfrac{j-1}{j+1}\right)^2$.

   Answer: $\dfrac{49}{36}$

   Difficulty: 1   Section: 1

10. Write the sum $9 + 7 + 5 + 3 + 1 - 1 - 3$ in sigma notation.

    Answer: $\sum\limits_{i=1}^{7} (11 - 2i)$

    Difficulty: 2   Section: 1

11. Write the sum $81 - 27 + 9 - 3 + 1$ in sigma notation.

    Answer: $\sum\limits_{k=0}^{4} (-1)^k 3^{4-k}$

    Difficulty: 2   Section: 1

12. Write the sum $-1^2 - 2^2 - 3^2 - 4^2 - 5^2 - 6^2$ in sigma notation.

    Answer: $\sum\limits_{m=1}^{6} (-m^2)$

    Difficulty: 2   Section: 1

13. Find the value of $\sum\limits_{k=1}^{50} \left[(k+1)^2 - k^2\right]$.

    Answer: 2600

    Difficulty: 2   Section: 1

14. Evaluate the sum $\sum\limits_{k=1}^{n} (6k^2 + 3)$.

    Answer: $2n^3 + 3n^2 + 4n$

    Difficulty: 2   Section: 1

15. Evaluate the sum $\sum\limits_{j=1}^{n} (3j^2 - 2j + 1)$.

    Answer: $\dfrac{1}{2}(2n^3 + n^2 + n)$

    Difficulty: 2   Section: 1

16. Evaluate the sum $\sum_{t=1}^{n} (t^2 - 2t)$.

    Answer: $\dfrac{1}{6}\left(2n^3 - 3n^2 - 5n\right)$

    Difficulty: 2   Section: 1

17. Evaluate the sum $\sum_{k=2}^{n} (2k - 3)$.

    Answer: $n^2 - 2n + 1$

    Difficulty: 2   Section: 1

18. Find the area under the curve $y = 4x - 3$ over the interval $[1, 4]$. To do this, divide the interval into n equal subintervals, calculate the area of the corresponding circumscribed polygon and then let $n \to \infty$.

    Answer: 21 square units

    Difficulty: 1   Section: 2

19. Find the area under the curve $y = 2x + 5$ over the interval $[1, 5]$. To do this, divide the interval into n equal subintervals, calculate the area of the corresponding inscribed polygon and then let $n \to \infty$.

    Answer: 44 square units

    Difficulty: 1   Section: 2

20. Find the area under the curve $y = x^2 + 2$ over the interval $[0, 3]$. To do this, divide the interval into n equal subintervals, calculate the area of the corresponding inscribed polygon and then let $n \to \infty$.

    Answer: 15 square units

    Difficulty: 1   Section: 2

21. Find the area under the curve $y = x^3$ over the interval $[0, 2]$. To do this, divide the interval into n equal subintervals, calculate the area of the corresponding circumscribed polygon and then let $n \to \infty$.

    Answer: 4 square units

    Difficulty: 1   Section: 2

22. Find the area under the curve $y = 3 + x^3$ over the interval $[0, 2]$. To do this, divide the interval into n equal subintervals, calculate the area of the corresponding circumscribed polygon and then let $n \to \infty$.

    Answer: 10 square units

    Difficulty: 2   Section: 2

23. Use geometry to calculate $\int_1^3 (2x+3)\, dx$.

    Answer: 14

    Difficulty: 1   Section: 2

24. Use geometry to calculate $\int_1^2 (2x-1)\, dx$.

    Answer: 2

    Difficulty: 1   Section: 2

25. Use geometry to calculate $\int_{-1}^2 (2+4x)\, dx$.

    Answer: 12

    Difficulty: 1   Section: 2

26. Evaluate $\int_1^2 (3-x)\, dx$ using the definition of the definite integral.

    Answer: $\dfrac{3}{2}$

    Difficulty: 2   Section: 2

27. Evaluate $\int_2^3 (x^2+6)\, dx$ using the definition of the definite integral.

    Answer: $\dfrac{37}{3}$

    Difficulty: 2   Section: 2

28. Evaluate $\int_0^3 (x^2+2)\, dx$ using the definition of the definite integral.

    Answer: 15

    Difficulty: 2   Section: 2

29. Evaluate $\int_3^3 \left(\cos^2 2x + \sqrt{x}\right) dx$.

    Answer: 0

    Difficulty: 1   Section: 2

30. Evaluate $\int_4^1 x\, dx$.

    Answer: $-\dfrac{15}{2}$

Difficulty: 1   Section: 2

31. If $\int_{-2}^{1} f(x)\,dx = 4$ and $\int_{0}^{1} f(x)\,dx = 2$, find $\int_{-2}^{0} f(x)\,dx$.

   Answer: 2

   Difficulty: 1   Section: 2

32. If $\int_{0}^{2} f(x)\,dx = 3$ and $\int_{-3}^{0} f(x)\,dx = -3$, find $\int_{-3}^{2} f(x)\,dx$.

   Answer: 0

   Difficulty: 1   Section: 2

33. Find $D_x \left[\int_{0}^{x} \sin(5w)\,dw \right]$.

   Answer: $\sin 5x$

   Difficulty: 1   Section: 3

34. Find $D_z \left[\int_{-1}^{z} (2x-5)^5\,dx \right]$.

   Answer: $(2z-5)^5$

   Difficulty: 1   Section: 3

35. Find $D_x \left[\int_{x}^{0} (2p^3 - 7)\,dp \right]$.

   Answer: $7 - 2p^3$

   Difficulty: 1   Section: 3

36. Find $\dfrac{d}{dy} \left[\int_{y}^{-1} (\cos 5x - 3x^2)\,dx \right]$.

   Answer: $3y^2 - \cos 5y$

   Difficulty: 1   Section: 3

37. Find $\dfrac{d}{dx} \left[\int_{y}^{x^3} \cos(3w)\,dw \right]$.

   Answer: $3x^2 \cos(3x^3)$

   Difficulty: 2   Section: 3

38. Find $\dfrac{d}{dx} \left[\int_{0}^{\sin x} (2t^2 + t)\,dt \right]$.

Answer: $(2\sin^2 x + \sin x) \cdot \cos x$

Difficulty: 2   Section: 3

39. Find $\dfrac{d}{dy}\left[\displaystyle\int_{-1}^{3y} (1 - 3t^2)\, dt\right]$.

Answer: $3(1 - 27y^2)$

Difficulty: 2   Section: 3

40. Use the Second Fundamental Theorem of Calculus to evaluate $\displaystyle\int_1^2 (2t - 1)\, dt$.

Answer: 2

Difficulty: 1   Section: 4

41. Use the Second Fundamental Theorem of Calculus to evaluate $\displaystyle\int_1^2 (3 - x)\, dx$.

Answer: $\dfrac{3}{2}$

Difficulty: 1   Section: 4

42. Use the Second Fundamental Theorem of Calculus to evaluate $\displaystyle\int_1^4 \pi x\, dx$.

Answer: $\dfrac{15\pi}{2}$

Difficulty: 1   Section: 4

43. Evaluate $\displaystyle\int_3^{-1} 4x^3\, dx$.

Answer: -80

Difficulty: 1   Section: 4

44. Use the Second Fundamental Theorem of Calculus to evaluate $\displaystyle\int_{-1}^2 (2 + 4t)\, dt$.

Answer: 12

Difficulty: 1   Section: 4

45. Use the Second Fundamental Theorem of Calculus to evaluate $\displaystyle\int_1^9 \sqrt{x}\, dx$.

Answer: $\dfrac{52}{3}$

Difficulty: 1   Section: 4

46. Use the Second Fundamental Theorem of Calculus to evaluate $\int_2^3 \frac{4}{x^3}\,dx$.

   Answer: $\frac{5}{18}$

   Difficulty: 1   Section: 4

47. Use the Second Fundamental Theorem of Calculus to evaluate $\int_1^2 \frac{x^3+1}{x^2}\,dx$.

   Answer: 2

   Difficulty: 2   Section: 4

48. Use the Second Fundamental Theorem of Calculus to evaluate $\int_{-2}^2 \left(4-x^2\right)\,dx$.

   Answer: $\frac{32}{3}$

   Difficulty: 2   Section: 4

49. Use the Second Fundamental Theorem of Calculus to evaluate $\int_{-1}^1 \left(x^2-3x\right)\,dx$.

   Answer: $\frac{2}{3}$

   Difficulty: 2   Section: 4

50. Use the Second Fundamental Theorem of Calculus to evaluate $\int_{-1}^3 \left(6x^2-1\right)\,dx$.

   Answer: 52

   Difficulty: 2   Section: 4

51. Use the Second Fundamental Theorem of Calculus to evaluate $\int_2^3 \left(x^2+6\right)\,dx$.

   Answer: $\frac{37}{3}$

   Difficulty: 2   Section: 4

52. Use the Second Fundamental Theorem of Calculus to evaluate $\int_{-1}^1 \left(x^2-1\right)^2\,dx$.

   Answer: $\frac{16}{15}$

   Difficulty: 2   Section: 4

53. Use the Second Fundamental Theorem of Calculus to evaluate $\int_0^1 \left(z^2+1\right)^2\,dz$.

Answer: $\dfrac{28}{15}$

Difficulty: 2   Section: 4

54. Use the Second Fundamental Theorem of Calculus to evaluate $\int_{-1}^{2}(w+2-w^2)\,dw$.

Answer: $\dfrac{9}{2}$

Difficulty: 2   Section: 4

55. Use the Second Fundamental Theorem of Calculus to evaluate $\int_{0}^{\pi/6}\cos 2x\,dx$.

Answer: $\dfrac{\sqrt{3}}{4}$

Difficulty: 2   Section: 4

56. Evaluate $\int_{1}^{2}2\sqrt{2-x}\,dx$.

Answer: $\dfrac{4}{3}$

Difficulty: 2   Section: 4

57. Evaluate $\int_{0}^{2}x^2\sqrt{x^3+1}\,dx$.

Answer: $\dfrac{52}{9}$

Difficulty: 2   Section: 4

58. Evaluate $\int_{2}^{3}x\sqrt{x^2-4}\,dx$.

Answer: $\dfrac{5\sqrt{5}}{3}$

Difficulty: 2   Section: 4

59. Evaluate $\int_{0}^{2}x^2\sqrt{x^3+1}\,dx$.

Answer: $\dfrac{52}{9}$

Difficulty: 2   Section: 4

60. Find $\int\left(1-\dfrac{3}{x}\right)^5\dfrac{1}{x^2}\,dx$.

Answer: $\dfrac{1}{18}\left(1-\dfrac{3}{x}\right)^6 + C$

Difficulty: 2   Section: 4

61. Find $\displaystyle\int \dfrac{5}{(7x-2)^2}\, dx$.

Answer: $-\dfrac{5}{7(7x-2)} + C$

Difficulty: 1   Section: 4

62. Find $\displaystyle\int \dfrac{x}{\sqrt[3]{1+x^2}}\, dx$.

Answer: $\dfrac{3}{4}\left(1+x^2\right)^{2/3} + C$

Difficulty: 2   Section: 4

63. Find $\displaystyle\int \dfrac{x}{\sqrt[5]{1-x^2}}\, dx$.

Answer: $-\dfrac{5}{8}\left(1-x^2\right)^{4/5} + C$

Difficulty: 2   Section: 4

64. Find $\displaystyle\int \dfrac{1}{\sqrt{x}\left(1+\sqrt{x}\right)^2}\, dx$.

Answer: $-\dfrac{2}{1+\sqrt{x}} + C$

Difficulty: 2   Section: 4

65. Find $\displaystyle\int \dfrac{x-2}{(x^2-4x+3)^3}\, dx$.

Answer: $-\dfrac{1}{4(x^2-4x+3)^2} + C$

Difficulty: 2   Section: 4

66. Find $\displaystyle\int (3x+2)\sqrt{3x^2+4x+9}\, dx$.

Answer: $\dfrac{1}{3}\left(3x^2+4x+9\right)^{3/2} + C$

Difficulty: 2   Section: 4

67. Find $\displaystyle\int \dfrac{\sqrt[4]{1-x^{-1}}}{x^2}\, dx$.

Answer: $\dfrac{4}{5}\left(1-x^{-1}\right)^{5/4}+C$

Difficulty: 2 Section: 4

68. Find $\displaystyle\int \cos^2 x \sin x\, dx$.

Answer: $-\dfrac{1}{3}\cos^3 x + C$

Difficulty: 2 Section: 4

69. Find $\displaystyle\int 3t\left[\sec^2\left(2t^2\right)\right]\, dt$.

Answer: $\dfrac{3}{4}\tan\left(2t^2\right) + C$

Difficulty: 2 Section: 4

70. Find $\displaystyle\int \sec 3x \tan 3x\, dx$.

Answer: $\dfrac{1}{3}\sec 3x + C$

Difficulty: 2 Section: 4

71. Find $\displaystyle\int \left[\sec^2 2x + x\sec\left(x^2\right)\tan\left(x^2\right)\right]\, dx$.

Answer: $\dfrac{1}{2}\left[\tan 2x + \sec\left(x^2\right)\right] + C$

Difficulty: 2 Section: 4

72. Find $\displaystyle\int \dfrac{1}{2}\sec^2\left(\dfrac{3x}{2}\right)\, dx$.

Answer: $\dfrac{1}{3}\tan\left(\dfrac{3x}{2}\right) + C$

Difficulty: 2 Section: 4

73. Find $\displaystyle\int \dfrac{\cos x}{\sqrt{\sin x}}\, dx$.

Answer: $2\sqrt{\sin x} + C$

Difficulty: 2 Section: 4

74. Find $\displaystyle\int \dfrac{\cos x}{(1-\sin x)^2}\, dx$.

Answer: $\dfrac{1}{1-\sin x} + C$

Difficulty: 2   Section: 4

75. Find $\int \cos^2 2x \sin 2x \, dx$.

Answer: $-\dfrac{1}{6}\cos^3 2x + C$

Difficulty: 2   Section: 4

76. Find $\int \sqrt{\dfrac{\sin x}{\cos^5 x}}\, dx$.

Answer: $\dfrac{2}{3}\tan^{3/2} x + C$

Difficulty: 2   Section: 4

77. Find $\int \dfrac{\cos 3\sqrt{x}}{\sqrt{x}}\, dx$.

Answer: $\dfrac{2}{3}\sin 3\sqrt{x} + C$

Difficulty: 2   Section: 4

78. Find $\int \sin^2 x \, dx$ given that $\sin^2 x = \dfrac{1}{2}(1 - \cos 2x)$.

Answer: $\dfrac{1}{2}x - \dfrac{1}{4}\sin 2x + C$

Difficulty: 2   Section: 4

79. Let $f(x) = \cos 2x$. Find an antiderivative whose graph passes through $\left(\dfrac{\pi}{6}, 1\right)$.

Answer: $\dfrac{1}{2}\sin 2x + \dfrac{4 - \sqrt{3}}{4}$

Difficulty: 1   Section: 4

80. Let $f(x) = \cos 3x - \sin 3x$. Find an antiderivative whose graph passes through $\left(\pi, \dfrac{5}{3}\right)$.

Answer: $\dfrac{1}{3}(\sin 3x + \cos 3x) + 2$

Difficulty: 1   Section: 4

81. Let $f(x) = \sec^2 \dfrac{x}{2}$. Find an antiderivative whose graph passes through $\left(\dfrac{\pi}{2}, 0\right)$.

Answer: $2\tan \dfrac{x}{2} - 2$

Difficulty: 1   Section: 4

82. Let $f(x) = x^2\sqrt{4 + 5x^3}$. Find an antiderivative whose graph passes through $(0, 1)$.

Answer: $\dfrac{2}{45}(4+5x^3)^{3/2} + \dfrac{29}{45}$

Difficulty: 2 Section: 4

83. Let $f(x) = (\cos x)\sqrt{4+5\sin x}$. Find an antiderivative whose graph passes through $\left(\dfrac{\pi}{2}, \dfrac{18}{5}\right)$.

Answer: $\dfrac{2}{15}(4+5\sin x)^{3/2}$

Difficulty: 2 Section: 4

84. Let $f(x) = x^3\sqrt{1-x^2}$. Find an antiderivative whose graph passes through $(0,5)$.

Answer: $-\dfrac{3}{8}(1-x^2)^{4/3} + \dfrac{43}{8}$

Difficulty: 2 Section: 4

85. A particle is moving along a line with acceleration $a(t) = (2t+1)^6$. Find the velocity function if $v(0) = 0$.

Answer: $\dfrac{1}{14}(2t+1)^7 - \dfrac{1}{14}$

Difficulty: 2 Section: 4

86. A particle is moving along a line with velocity $v(t) = t\cos(t^2)$. Find the position function if $s(\sqrt{\pi}) = \pi$.

Answer: $\dfrac{1}{2}\sin(t^2) + \pi$

Difficulty: 2 Section: 4

87. A particle is moving along a line with velocity $v(t) = \sin 2t + \cos 2t$. Find the position function $s(t)$ if $s\left(\dfrac{\pi}{4}\right) = 2$.

Answer: $\dfrac{1}{2}(\sin 2t - \cos 2t) + \dfrac{3}{2}$

Difficulty: 2 Section: 4

88. Evaluate $\displaystyle\int_0^{4/3} \sqrt{1+\dfrac{9}{4}x}\,dx$.

Answer: $\dfrac{56}{27}$

Difficulty: 1 Section: 4

89. Evaluate $\displaystyle\int_0^1 x^2\sqrt{4+5x^3}\,dx$.

Answer: $\dfrac{38}{45}$

Difficulty: 1   Section: 4

90. Evaluate $\int_0^2 \dfrac{1}{(3x+2)^2}\,dx$.

    Answer: $\dfrac{1}{8}$

    Difficulty: 2   Section: 4

91. Evaluate $\int_4^8 \dfrac{x}{\sqrt{x^2-15}}\,dx$.

    Answer: 6

    Difficulty: 2   Section: 4

92. Evaluate $\int_0^2 \dfrac{4y}{\sqrt{25-4y^2}}\,dx$.

    Answer: 2

    Difficulty: 2   Section: 4

93. Evaluate $\int_0^1 18x\sqrt{3x^2+1}\,dx$.

    Answer: 14

    Difficulty: 2   Section: 4

94. Evaluate $\int_0^{\sqrt{2}/4} \dfrac{t}{\sqrt{1-4t^2}}\,dt$.

    Answer: $\dfrac{1}{4}\left(1-\dfrac{\sqrt{2}}{2}\right)$

    Difficulty: 2   Section: 4

95. Evaluate $\int_4^8 \dfrac{3t}{\sqrt{t^2-15}}\,dt$.

    Answer: 18

    Difficulty: 2   Section: 4

96. Evaluate $\int_0^1 (3x+2)\left(3x^2+4x+9\right)^{1/2}\,dx$.

    Answer: $\dfrac{37}{3}$

    Difficulty: 2   Section: 4

97. Evaluate $\int_{\pi/18}^{\pi/9} \sin 3w \, dw$.

   Answer: $\dfrac{\sqrt{3}-1}{6}$

   Difficulty: 1   Section: 4

98. Evaluate $\int_0^{\pi/2} \cos(6p) \, dp$.

   Answer: 0

   Difficulty: 1   Section: 4

99. Evaluate $\int_0^{\pi/2} \sec^2 \dfrac{x}{2} \, dx$.

   Answer: 2

   Difficulty: 1   Section: 4

100. Evaluate $\int_{\pi/6}^{\pi/2} \cos^2 x \sin x \, dx$.

   Answer: $\dfrac{\sqrt{3}}{8}$

   Difficulty: 1   Section: 4

101. Evaluate $\int_0^{\pi/4} \cos^2 2x \sin 2x \, dx$.

   Answer: $\dfrac{1}{6}$

   Difficulty: 1   Section: 4

102. Evaluate $\int_{\pi/2}^{\pi} \sin^3 \theta \cos \theta \, d\theta$.

   Answer: $-\dfrac{1}{4}$

   Difficulty: 1   Section: 4

103. Evaluate $\int_{\pi/4}^{\pi/3} (1 + \cos t) \sin t \, dt$.

   Answer: $\dfrac{1}{8}\left(4\sqrt{2} - 3\right)$

   Difficulty: 1   Section: 4

104. Evaluate $\int_0^{\pi/2} \dfrac{\cos x}{(2 - \sin x)^{1/3}} \, dx$.

Answer: $\dfrac{3}{2}\left(\sqrt[3]{4}-1\right)$

Difficulty: 2   Section: 4

105. Find all values of c that satisfy the Mean Value Theorem for Integrals for $f(x)=\sqrt{x+4}$ on the interval $[0,12]$.

Answer: $\dfrac{460}{81}$

Difficulty: 1   Section 5

106. Find all values of c that satisfy the Mean Value Theorem for Integrals for $f(x)=x^4$ on the interval $[-2,2]$.

Answer: $\pm\dfrac{2}{5^{1/4}}$

Difficulty: 1   Section 5

107. Find all values of c that satisfy the Mean Value Theorem for Integrals for $f(x)=\cos 2x$ on the interval $[0,\pi]$.

Answer: $\dfrac{\pi}{4}, \dfrac{3\pi}{4}$

Difficulty: 1   Section 5

108. Use the Trapezoid Rule, with $n=4$, to approximate $\displaystyle\int_0^1 \dfrac{1}{(x+1)^2}\,dx$. Round your answer to two decimal places.

Answer: 0.51

Difficulty: 1   Section: 6

109. What value of n should be used with the Trapezoid Rule to approximate $\displaystyle\int_1^4 \sqrt{x}\,dx$ to within 0.01?

Answer: $n \geq 8$

Difficulty: 2   Section: 6

110. What value of n should be used with the Trapezoid Rule to approximate $\displaystyle\int_1^2 \dfrac{1}{t}\,dt$ to within 0.00001?

Answer: $n \geq 41$

Difficulty: 2   Section: 6

111. Use Simpson's Rule, with $n=8$, to approximate $\displaystyle\int_0^1 \dfrac{1}{1+x^2}\,dx$. Round your answer to 4 decimal places.

Answer: 0.7854

Difficulty: 1 Section: 6

112. Use Simpson's Rule, with $n = 4$, to approximate $\int_0^1 \dfrac{1}{\sqrt{1+x^2}}\, dx$. Round your answer to three decimal places.

Answer: 0.881

Difficulty: 1 Section: 6

113. What value of n should be used in Simpson's Rule to approximate $\int_1^4 \sqrt{x}\, dx$ to within 0.01?

Answer: $n \geq 4$

Difficulty: 2 Section: 6

114. What value of n should be used in Simpson's Rule to approximate $\int_1^2 \dfrac{1}{x}\, dx$ to within 0.0001?

Answer: $n \geq 8$

Difficulty: 2 Section: 6

5 Applications of the Integral

1. Find the area of the region bounded by the x-axis, the graph of $f(x) = x^2$, $x = 0$, and $x = 1$.

   Answer: $\dfrac{1}{3}$

   Difficulty: 1   Section: 1

2. Find the area of the region bounded by the x-axis, the graph of $f(x) = 4x - 3$, $x = 1$, and $x = 9$.

   Answer: 21

   Difficulty: 1   Section: 1

3. Find the area of the region bounded by the x-axis, the graph of $f(x) = x^3 + 3$, $x = 0$, and $x = 2$.

   Answer: 10

   Difficulty: 1   Section: 1

4. Find the area of the region bounded by the x-axis, the graph of $f(x) = x^2 - 4x + 5$, $x = 1$, and $x = 3$.

   Answer: $\dfrac{8}{3}$

   Difficulty: 1   Section: 1

5. Find the area of the region bounded by the x-axis, the graph of $f(x) = -x^2 + x + 1$, $x = 0$, and $x = 3$.

   Answer: $\dfrac{7}{6}$

   Difficulty: 1   Section: 1

6. Find the area of the region bounded by the x-axis, the graph of $f(x) = \sin x$, $x = 0$, and $x = \pi$.

   Answer: 2

   Difficulty: 1   Section: 1

7. Find the area between the curves $y = x^2$ and $y = 2x^2 - 4$.

   Answer: $\dfrac{32}{3}$

   Difficulty: 1   Section: 1

8. Find the area between the curves $y = x^2$ and $y = 2 - x^2$.

   Answer: $\dfrac{8}{3}$

Difficulty: 1 Section: 1

9. Find the area between the curve $x = \cos^2 3y$ and the y-axis for $0 \leq y \leq \dfrac{\pi}{4}$.

   Answer: $\dfrac{\pi}{8} - \dfrac{1}{12}$

   Difficulty: 1 Section: 1

10. Find the area bounded by $y = x^2$ and $y = x + 2$.

    Answer: $\dfrac{9}{2}$

    Difficulty: 1 Section: 1

11. Find the area bounded by $y = x^2$ and $y = 4$.

    Answer: $\dfrac{32}{3}$

    Difficulty: 1 Section: 1

12. Find the area bounded by $y = x^2$ and $y = 2x - x^2$.

    Answer: $\dfrac{1}{3}$

    Difficulty: 1 Section: 1

13. Find the area bounded by $y = x^3 - x$ and the x-axis.

    Answer: $\dfrac{1}{2}$

    Difficulty: 1 Section: 1

14. Find the area bounded by $y = x^3 - 4x$ and the x-axis.

    Answer: 8

    Difficulty: 1 Section: 1

15. Find the area bounded by $x = y^2$ and $x = 2y - y^2$.

    Answer: $\dfrac{1}{3}$

    Difficulty: 2 Section: 1

16. Find the area bounded by $x = y - y^3$ and the y-axis.

    Answer: $\dfrac{1}{2}$

    Difficulty: 2 Section: 1

17. Interpret $\int_{-2}^{2} \sqrt{4-x^2}\,dx$ as an area and evaluate the definite integral without finding an antiderivative.

    Answer: The integral gives the area of a semicircle of radius 2 and the value is 2π.

    Difficulty: 2   Section: 1

18. Interpret $\int_{-3}^{3} \sqrt{9-x^2}\,dx$ as an area and evaluate the definite integral without finding an antiderivative.

    Answer: The integral gives the area of a semicircle of radius 3 and the value is $\dfrac{9}{2}\pi$.

    Difficulty: 2   Section: 1

19. Find the area enclosed by the ellipse $\dfrac{x^2}{a^2} + \dfrac{y^2}{b^2} = 1$.

    Answer: πab

    Difficulty: 3   Section: 1

20. The base of a solid is the region bounded by $y = x^2$ and $y = x$. A cross section by a plane perpendicular to the x-axis is a square, one side of which is in the base. Find the volume.

    Answer: $\dfrac{1}{30}$

    Difficulty: 1   Section: 2

21. The base of a solid is the region bounded by $y = x^2$ and $y = 2x$. A cross section by a plane perpendicular to the x-axis is a square, one side of which is in the base. Find the volume.

    Answer: $\dfrac{16}{15}$

    Difficulty: 1   Section: 2

22. The base of a solid is the region bounded by $y = x^2$ and $y = 3x$. A cross section by a plane perpendicular to the x-axis is a square, one side of which is in the base. Find the volume.

    Answer: $\dfrac{81}{10}$

    Difficulty: 1   Section: 2

23. The base of a solid is the region bounded by the ellipse $\dfrac{x^2}{4} + y^2 = 1$. A cross section by a plane perpendicular to the x-axis is a circle with diameter in the base. Find the volume.

    Answer: $\dfrac{8\pi}{3}$

    Difficulty: 1   Section: 2

24. The base of a solid is the region bounded by the ellipse $x^2 + \dfrac{y^2}{4} = 1$. A cross section by a plane

perpendicular to the x-axis is a circle with diameter in the base. Find the volume.

Answer: $\dfrac{16\pi}{3}$

Difficulty: 1   Section: 2

25. The base of a solid is the region bounded by the ellipse $\dfrac{x^2}{9} + \dfrac{y^2}{4} = 1$. A cross section by a plane perpendicular to the x-axis is a circle with diameter in the base. Find the volume.

Answer: 16π

Difficulty: 1   Section: 2

26. The base of a solid is the region bounded by $y = x$ and $y = x^2$. A cross section by a plane perpendicular to the x-axis is a circle with diameter in the base. Find the volume.

Answer: $\dfrac{\pi}{120}$

Difficulty: 1   Section: 2

27. The base of a solid is the region bounded by $y = 2x$ and $y = x^2$. A cross section by a plane perpendicular to the x-axis is a circle with diameter in the base. Find the volume.

Answer: $\dfrac{4\pi}{15}$

Difficulty: 1   Section: 2

28. The base of a solid is the region bounded by $y = 3x$ and $y = x^2$. A cross section by a plane perpendicular to the x-axis is a circle with diameter in the base. Find the volume.

Answer: $\dfrac{81\pi}{40}$

Difficulty: 1   Section: 2

29. The base of a solid is the region inside the ellipse $x^2 + \dfrac{y^2}{4} = 1$. A cross section by a plane perpendicular to the x-axis is a square with one side in the base. Find the volume.

Answer: $\dfrac{64}{3}$

Difficulty: 1   Section: 2

30. The base of a solid is the region inside the ellipse $\dfrac{x^2}{9} + \dfrac{y^2}{4} = 1$. A cross section by a plane perpendicular to the x-axis is a circle with diameter in the base. Find the volume.

Answer: 64

Difficulty: 1   Section: 2

31. The region bounded by $y = \sqrt{1 + x^2}$, the x-axis, the line $x = 1$ and the line $x = 2$ is rotated about the x-axis to form a solid. Find the volume.

Answer: $\dfrac{10\pi}{3}$

Difficulty: 1   Section: 2

32. The region bounded by $y = \sqrt{1+x^2}$, the x-axis, the line $x = 1$ and the line $x = 3$ is rotated about the x-axis to form a solid. Find the volume.

Answer: $\dfrac{32\pi}{3}$

Difficulty: 1   Section: 2

33. The region bounded by $y = \sqrt{1+x^2}$, the x-axis, the line $x = 1$ and the line $x = 2$ is rotated about the y-axis to form a solid. Find the volume.

Answer: $\dfrac{2\pi}{3}\left(5^{3/2} - 2^{3/2}\right)$

Difficulty: 1   Section: 2

34. The region bounded by $y = \sqrt{2+x^2}$, the x-axis, the line $x = 1$ and the line $x = 2$ is rotated about the x-axis to form a solid. Find the volume.

Answer: $\dfrac{13\pi}{3}$

Difficulty: 1   Section: 2

35. The region bounded by $y = \sqrt{2+x^2}$, the x-axis, the line $x = 1$ and the line $x = 2$ is rotated about the y-axis to form a solid. Find the volume.

Answer: $\dfrac{2\pi}{3}\left(6^{3/2} - 3^{3/2}\right)$

Difficulty: 1   Section: 2

36. Find the volume of the solid formed by revolving the region bounded by $y = x^2$, $x = 3$, and $y = 0$ about the y-axis.

Answer: $\dfrac{81\pi}{2}$

Difficulty: 1   Section: 2

37. Find the volume of the solid formed by revolving the region bounded by $y = x^2 - 2x$ and the x-axis about the line $y = 3$.

Answer: $\dfrac{136\pi}{15}$

Difficulty: 1   Section: 2

38. Find the volume of the solid formed by revolving the region bounded by $\dfrac{x^2}{9} + \dfrac{y^2}{4} = 1$ about the x-axis.

Answer: 16π

Difficulty: 1   Section: 2

39. Find the volume of the solid formed by revolving the region bounded by $x^2 + \dfrac{y^2}{4} = 1$ about the y-axis.

    Answer: $\dfrac{8\pi}{3}$

    Difficulty: 1   Section: 2

40. Find the volume of the solid formed by revolving the region bounded by $y = \sqrt{x}$, $x = 9$, and the x-axis about the y-axis.

    Answer: $\dfrac{972\pi}{5}$

    Difficulty: 1   Section: 2

41. Find the volume of the solid formed by revolving the region bounded by $y = \sqrt{x}$, $x = 9$, and the x-axis about the x-axis.

    Answer: $\dfrac{81\pi}{2}$

    Difficulty: 1   Section: 2

42. Find the volume of the solid formed by revolving the region bounded by $x^2 + \dfrac{y^2}{4} = 1$ about the x-axis.

    Answer: $\dfrac{16\pi}{3}$

    Difficulty: 1   Section: 2

43. The region bounded by $y = x$, $y = x^2$, $x = 1$, and $x = 2$ is rotated about the y-axis to form a solid. Use shells to find the volume.

    Answer: $\dfrac{17\pi}{6}$

    Difficulty: 1   Section: 3

44. The region bounded by $y = x$, $y = x^2$, $x = 1$, and $x = 2$ is rotated about the line $x = 2$. Use shells to find the volume.

    Answer: $\dfrac{\pi}{2}$

    Difficulty: 1   Section: 3

45. The region bounded by $y = x$, $y = x^2$, $x = 1$, and $x = 2$ is rotated about the line $x = 1$. Use shells to find the volume.

    Answer: $\dfrac{7\pi}{6}$

    Difficulty: 1   Section: 3

46. The region bounded by $y = \sqrt{1+x^2}$, the x-axis, the line $x = 1$, and the line $x = 2$ is revolved about the y-axis to form a solid. Use shells to find the volume.

   Answer: $\dfrac{2\pi}{3}\left(5^{3/2} - 2^{3/2}\right)$

   Difficulty: 1   Section: 3

47. The region bounded by $y = \sqrt{2+x^2}$, the x-axis, the line $x = 1$, and the line $x = 2$ is revolved about the y-axis to form a solid. Use shells to find the volume.

   Answer: $\dfrac{2\pi}{3}\left(6^{3/2} - 3^{3/2}\right)$

   Difficulty: 1   Section: 3

48. The region bounded by $y = x^2 - 2x$ and the x-axis is revolved about the line $x = 2$. Use shells to find the volume.

   Answer: $\dfrac{8\pi}{3}$

   Difficulty: 1   Section: 3

49. The region bounded by $y = x^2 - 2x$ and the x-axis is revolved about the line $x = 3$. Use shells to find the volume.

   Answer: $\dfrac{16\pi}{3}$

   Difficulty: 1   Section: 3

50. The region bounded by $y = x^2$ and $y = x + 2$ is revolved about the line $x = 2$. Use shells to find the volume.

   Answer: $\dfrac{27\pi}{2}$

   Difficulty: 1   Section: 3

51. The region bounded by $y = x^2$ and $y = x + 2$ is revolved about the line $x = -1$. Use shells to find the volume.

   Answer: $\dfrac{27\pi}{2}$

   Difficulty: 1   Section: 3

52. The region bounded by $y = 2x^2$ and the line $y = 2$ is revolved about the line $x = 1$. Use shells to find the volume.

   Answer: $\dfrac{16\pi}{3}$

   Difficulty: 1   Section: 3

53. The region bounded by $y = 2x^2$ and the line $y = 2$ is revolved about the line $x = 2$. Use shells to find the volume.

Answer: $\dfrac{32\pi}{3}$

Difficulty: 1   Section: 3

54. Write the integral whose value is the length of the curve $y = x^2$ from $x = 1$ to $x = 4$.

Answer: $\displaystyle\int_1^4 \sqrt{1 + 4x^2}\, dx$

Difficulty: 1   Section: 4

55. Write the integral whose value is the length of the curve $y = x^3$ from $x = 0$ to $x = 4$.

Answer: $\displaystyle\int_0^4 \sqrt{1 + 9x^4}\, dx$

Difficulty: 1   Section: 4

56. Write the integral whose value is the length of the curve $y = \sec(\pi x)$ from $x = 0$ to $x = \dfrac{1}{4}$.

Answer: $\displaystyle\int_0^{1/4} \sqrt{1 + \pi^2 \sec^2(\pi x) \tan^2(\pi x)}\, dx$

Difficulty: 1   Section: 4

57. Write the integral whose value is the length of the curve $y = x - x^2$ from $x = 0$ to $x = 2$.

Answer: $\displaystyle\int_0^2 \sqrt{1 + (1 - 2x)^2}\, dx$

Difficulty: 1   Section: 4

58. Write the integral whose value is the length of the curve $x = 2y - y^3$ from $y = 1$ to $y = 6$.

Answer: $\displaystyle\int_1^6 \sqrt{1 + (2 - 3y^2)^2}\, dy$

Difficulty: 1   Section: 4

59. The curve $y = x^2$ from $x = 1$ to $x = 4$ is rotated about the x-axis to form a surface. Write the integral whose value is the area of that surface.

Answer: $2\pi \displaystyle\int_1^4 x^2 \sqrt{1 + 4x^2}\, dx$

Difficulty: 1   Section: 4

60. The curve $y = x^3$ from $x = 0$ to $x = 4$ is rotated about the x-axis to form a surface. Write the integral whose value is the area of that surface.

Answer: $2\pi \displaystyle\int_0^4 x^3 \sqrt{1 + 9x^4}\, dx$

Difficulty: 1   Section: 4

61. The curve $y = \sec x$ from $x = 0$ to $x = \dfrac{\pi}{4}$ is rotated about the x-axis to form a surface. Write the integral whose value is the area of that surface.

    Answer: $2\pi \displaystyle\int_0^{\pi/4} \sec x \sqrt{1 + \sec^2 x \tan^2 x}\, dx$

    Difficulty: 1   Section: 4

62. The curve $y = \sqrt{x}$ from $x = 0$ to $x = 9$ is rotated about the y-axis to form a surface. Write the integral whose value is the area of that surface.

    Answer: $2\pi \displaystyle\int_0^3 y^2 \sqrt{1 + 4y^2}\, dy$

    Difficulty: 1   Section: 4

63. The curve $y = \sqrt[3]{x}$ from $x = 0$ to $x = 8$ is rotated about the y-axis to form a surface. Write the integral whose value is the area of that surface.

    Answer: $2\pi \displaystyle\int_0^2 y^3 \sqrt{1 + 9y^4}\, dy$

    Difficulty: 1   Section: 4

64. The curve $y = \sqrt[3]{x^2}$ from $x = 1$ to $x = 8$ is rotated about the y-axis to form a surface. Write the integral whose value is the area of that surface.

    Answer: $2\pi \displaystyle\int_1^4 y^{3/2} \sqrt{1 + \dfrac{9}{4} y^{1/4}}\, dy$

    Difficulty: 1   Section: 4

65. Find the area of the surface generated by rotating the curve $x = t$, $y = t^3$ from $t = 1$ to $t = 2$ about the x-axis.

    Answer: $\dfrac{\pi}{27} \left(145^{3/2} - 10^{3/2} \right)$

    Difficulty: 2   Section: 4

66. Find the area of the surface generated by rotating the curve $x = 1 + t^2$ and $y = t$ from $t = 0$ to $t = 4$ about the x-axis.

    Answer: $\dfrac{\pi}{6} \left(65^{3/2} - 1 \right)$

    Difficulty: 2   Section: 4

67. Find the length of the curve given by $y = t^3$ and $x = t^2$ from $t = 0$ to $t = 1$.

    Answer: $\dfrac{1}{27} \left(13^{3/2} - 8 \right)$

    Difficulty: 2   Section: 4

68. A cylindrical water tank 6 feet in diameter and 8 feet long is lying on its side. If the tank is

full of water, how much work is done in pumping the water to a height one foot above the top of the tank? Use 62.5 as the density factor.

Answer: 288π (62.5) foot-pounds

Difficulty: 2   Section: 5

69. A cylindrical water tank 6 feet in diameter and 8 feet long is lying on its side. If the tank is half full of water, how much work is done in pumping the water over the top of the tank? . Use 62.5 as the density factor.

Answer: 216π (62.5) foot-pounds

Difficulty: 2   Section: 5

70. A water trough is 6 feet long and 2 feet wide at the top. The vertical cross section is in the shape of the parabola $y = x^2$ with the vertex down. If the trough is full, how much work is done in pumping the water over the top of the tank? Use 62.5 as the density factor.

Answer: $\dfrac{16}{5}$ (62.5) foot-pounds

Difficulty: 2   Section: 5

71. The ends of a water trough 6 feet long have the shape of isosceles trapezoids of lower base 2 feet, upper base 3 feet and altitude 2 feet. If the trapezoid is full, find the work done in pumping the water over the top of the tank. Use 62.5 as the density factor.

Answer: 28 (62.5) foot-pounds

Difficulty: 2   Section: 5

72. A spring has spring constant $k = 3.5$. Find the work done in stretching the spring 4 inches from its natural length.

Answer: 28 inch-pounds

Difficulty: 2   Section: 5

73. A spring has natural length of 10 inches. It takes a force of 4 pounds to stretch it to a length of 12 inches. How much work is done in stretching the spring to a length of 14 inches?

Answer: 16 inch-pounds

Difficulty: 2   Section: 5

74. The shaded region below is part of a vertical side of a tank full of water ($\delta = 62.4$ pounds per cubic foot). Find the total force exerted by the water against this region.

Answer: 4492.8 foot-pounds

Difficulty: 2   Section 5

75. The shaded region below is part of a vertical side of a tank full of water ($\delta = 62.4$ pounds per cubic foot). Find the total force exerted by the water against this region.

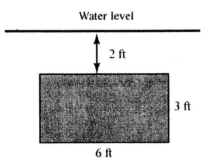

Answer: 3931.2 foot-pounds

Difficulty: 2   Section 5

76. The shaded region below is part of a vertical side of a tank full of water ($\delta = 62.4$ pounds per cubic foot). Find the total force exerted by the water against this region.

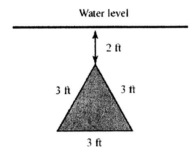

Answer: 907.56 foot-pounds

Difficulty: 2   Section 5

77. Find the centroid of the region bounded by $y = x + 2$ and $y = x^2$.

Answer: $\left(\dfrac{1}{2}, \dfrac{8}{5}\right)$

Difficulty: 1   Section: 6

78. Find the centroid of the upper half of the ellipse $x^2 + 4y^2 = 4$.

Answer: $\left(0, \dfrac{4}{3\pi}\right)$

Difficulty: 1   Section: 6

79. The centroid of the upper half of the ellipse $x^2 + 4y^2 = 4$ is $\left(0, \dfrac{4}{3\pi}\right)$. Use Pappus' Theorem to

find the volume of the solid formed by rotating the region about the x-axis.

Answer: $\dfrac{8\pi}{3}$

Difficulty: 2   Section: 6

80. Use Pappus' Theorem to find the centroid of the upper half of the circle $x^2 + y^2 = 9$. (Note: Rotating the region about the x-axis gives the sphere a radius of 3.)

Answer: $\dfrac{4}{\pi}$

Difficulty: 2   Section: 6

81. Find the centroid of the region bounded by $y = x^2$ and $y = 1$.

Answer: $\left(0, \dfrac{3}{5}\right)$

Difficulty: 1   Section: 6

82. Find the centroid of the triangle with vertices $(0,0)$, $(0,2)$, and $(3,0)$.

Answer: $\left(1, \dfrac{2}{3}\right)$

Difficulty: 1   Section: 6

83. An ellipse with major axis 4 and minor axis 2 is revolved about a line through a vertex on the major axis, perpendicular to the major axis. Find the volume of the solid thus formed.

Answer: $\dfrac{8\pi}{3}$

Difficulty: 2   Section: 6

84. A random variable X has probability density function
$$f(x) = \begin{cases} \dfrac{1}{36}x(6-x), & \text{if } 0 \le x \le 6 \\ 0, & \text{otherwise} \end{cases}$$
Find $P(X > 4)$ and $E(X)$.

Answer: $\dfrac{7}{27}$, 3

Difficulty: 1   Section: 7

85. A random variable X has probability density function
$$f(x) = \begin{cases} \dfrac{1}{800}(40-x), & \text{if } 0 \le x \le 40 \\ 0, & \text{otherwise} \end{cases}$$
Find $P(10 \le X < 20)$ and $E(X)$.

Answer: $\dfrac{5}{16}$, $\dfrac{40}{3}$

Difficulty: 1   Section: 7

6 Transcendental Functions

1. Use properties of $\ln(x)$ to simplify the function $\ln\left[\left(3x^3 + \sin 2x\right)^{1/3}\right]$.

   Answer: $\dfrac{1}{3}\ln\left(3x^3 + \sin 2x\right)$

   Difficulty: 1 Section: 1

2. Use properties of $\ln(x)$ to simplify the function $\ln\left[\dfrac{\sqrt{x}\cdot \cos 2x \cdot \tan x}{x^8\left(1+x^2\right)}\right]$.

   Answer: $\ln(\cos 2x) + \ln(\tan x) - \dfrac{15}{2}\ln(x) - \ln\left(1+x^2\right)$

   Difficulty: 1 Section: 1

3. Let $f(x) = \left(x^3 - 1\right)\ln|x|$. Find $f'(x)$.

   Answer: $\dfrac{x^3 - 1}{x} + 3x^2 \ln|x|$

   Difficulty: 2 Section: 1

4. Let $f(x) = \ln|6x - 7|$. Find $f'(x)$.

   Answer: $\dfrac{6}{6x - 7}$

   Difficulty: 1 Section: 1

5. Let $f(x) = \ln\left(1 + x^2\right)$. Find $\dfrac{dy}{dx}$.

   Answer: $\dfrac{2x}{1 + x^2}$

   Difficulty: 1 Section: 1

6. Let $f(x) = \ln\left|\dfrac{2x}{3x + 4}\right|$. Find $f'(x)$.

   Answer: $\dfrac{1}{x} - \dfrac{3}{3x + 4}$

   Difficulty: 2 Section: 1

7. Find the derivative of $[\ln|x|]^4$.

   Answer: $\dfrac{4}{x}[\ln|x|]^3$

   Difficulty: 2 Section: 1

8. Let $y = \ln|\sin x|$. Find $\dfrac{dy}{dx}$.

Answer: $\cot x$

Difficulty: 1   Section: 1

9. Let $f(x) = \ln(\cos^2 3x)$. Find $f'(x)$.

   Answer: $-6\tan 3x$

   Difficulty: 2   Section: 1

10. Evaluate $\int_1^4 \left(\sqrt{u} + \dfrac{1}{u}\right) du$.

    Answer: $\dfrac{14}{3} + \ln 4$

    Difficulty: 1   Section: 1

11. Evaluate $\int_0^1 \dfrac{2}{2-x^3}\, dx$.

    Answer: $\dfrac{\ln(2)}{3}$

    Difficulty: 1   Section: 1

12. Evaluate $\int_0^{\pi/4} \tan x\, dx$.

    Answer: $\ln(2\sqrt{2})$

    Difficulty: 2   Section: 1

13. Find $\int \dfrac{6x+1}{3x^2+x-19}\, dx$.

    Answer: $\ln|3x^2+x-19| + C$

    Difficulty: 2   Section: 1

14. Evaluate $\int \dfrac{1}{x[\ln(x)]^2}\, dx$.

    Answer: $-\dfrac{1}{\ln(x)} + C$

    Difficulty: 1   Section: 1

15. Let $f(x) = \dfrac{\sqrt{x^2+1}}{(9x-4)^2}$. Use logarithmic differentiation to find $f'(x)$.

    Answer: $\dfrac{\sqrt{x^2+1}}{(9x-4)^2}\left(\dfrac{x}{x^2+1} - \dfrac{18}{9x-4}\right)$

    Difficulty: 2   Section: 1

16. Let $f(x) = \sqrt{4x+7}(x-5)^3$. Use logarithmic differentiation to find $f'(x)$, then simplify.

    Answer: $\dfrac{(14x+11)(x-5)^2}{\sqrt{4x+7}}$

    Difficulty: 2   Section: 1

17. What is the slope of the tangent line to the graph of $3y - x^2 + \ln|xy| = 2$ at the point $(1, 1)$?

    Answer: $\dfrac{1}{4}$

    Difficulty: 2   Section: 1

18. Use implicit differentiation to find $\dfrac{dy}{dx}$ if $3y - x^2 + \ln|xy| + 5 = 0$.

    Answer: $\dfrac{y(2x^2 - 1)}{x(3y+1)}$

    Difficulty: 2   Section: 1

19. Write the equation of the tangent line to the graph of $y^3 - 4y = x(\ln x)$ at the point $(1, 2)$.

    Answer: $y - 2 = \dfrac{1}{8}(x - 1)$

    Difficulty: 2   Section: 1

20. Find the area of the region in the first quadrant that is bounded by the curves $y = x$, $y = \dfrac{2}{x} - 1$, and $y = 0$.

    Answer: $2\ln(2) - \dfrac{1}{2}$ square units

    Difficulty: 2   Section: 1

21. Let $f(x) = \dfrac{1}{3} - \dfrac{2x}{5}$. Find a formula for $f^{-1}(x)$.

    Answer: $f^{-1}(x) = \dfrac{5}{6} - \dfrac{5x}{2}$

    Difficulty: 1   Section: 2

22. Let $f(x) = 1 + \dfrac{x}{3}$. Find a formula for $f^{-1}(x)$.

    Answer: $f^{-1}(x) = 3x - 3$

    Difficulty: 1   Section: 2

23. Find the inverse function of $f(x) = 3(x+2)^2 - 4$, $x \geq -2$.

    Answer: $f^{-1}(x) = \sqrt{\dfrac{x+4}{3}} - 2$, $x \geq -4$

    Difficulty: 2   Section: 2

24. Find the inverse function of $f(x) = 2(x-1)^2 - 1$, $x \geq 1$.

   Answer: $f^{-1}(x) = \sqrt{\dfrac{x+1}{2}} + 1$, $x \geq -1$

   Difficulty: 2   Section: 2

25. Graph both the function $y = 2x + 1$ and its inverse of the same coordinate axes.

   Answer:

   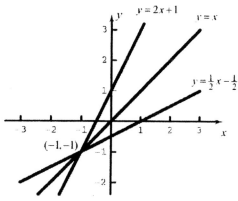

   Difficulty: 2   Section: 2

26. Graph both the function $y = -2x - 2$ and its inverse on the same coordinate axes.

   Answer:

   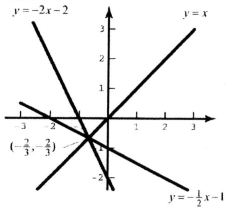

   Difficulty: 2   Section: 2

27. Let $f(x) = \dfrac{x}{x-2}$ and $F(x) = \dfrac{2x}{x-1}$. Apply the definition to show that $F(x) = f^{-1}(x)$.

   Answer: $F[f(x)] = F\left(\dfrac{x}{x-2}\right) = \dfrac{2\left(\dfrac{x}{x-2}\right)}{\left(\dfrac{x}{x-2}\right) - 1} = \dfrac{\dfrac{2x}{x-2}}{\dfrac{2}{x-2}} = x$

   Difficulty: 2   Section: 2

28. Let $f(x) = \dfrac{1}{x^3+1}$ and $F(x) = \sqrt[3]{\dfrac{1}{x}+1}$. Determine whether or not $F(x) = f^{-1}(x)$.

   Answer: It does not

   Difficulty: 2   Section: 2

29. Let $f(x) = \dfrac{x+2}{x-5}$ and $F(x) = \dfrac{5x-2}{x+1}$. Determine whether or not $F(x) = f^{-1}(x)$.

   Answer: It does not

   Difficulty: 2   Section: 2

30. Let $f(x) = \dfrac{x-1}{2x+1}$ and $F(x) = -\dfrac{x+1}{2x+1}$. Determine whether or not $F(x) = f^{-1}(x)$.

   Answer: In this case, $F(x) = f^{-1}(x)$

   Difficulty: 2   Section: 2

31. Let $f(x) = 4x+5$. Evaluate $(f^{-1})'(x)$ at $x = -10$.

   Answer: $\dfrac{1}{4}$

   Difficulty: 1   Section: 2

32. Let $f(x) = 2 - x^2$ for $x \geq 0$. Evaluate $(f^{-1})'(x)$ at $x = 1$.

   Answer: $-\dfrac{1}{2}$

   Difficulty: 1   Section: 2

33. Find the value of the expression $\ln \dfrac{1}{\sqrt[3]{e}}$.

   Answer: $-\dfrac{1}{3}$

   Difficulty: 1   Section: 3

34. Find the value of the expression $e^{\ln 1/2}$.

   Answer: $\dfrac{1}{2}$

   Difficulty: 1   Section: 3

35. Find the value of the expression $\left(\dfrac{1}{e}\right)^{-\ln 10^2}$.

   Answer: 100

   Difficulty: 1   Section: 3

36. Find the value of the expression $\dfrac{e^{\ln 7}}{e^{\ln 5}}$.

Answer: $\dfrac{7}{5}$

Difficulty: 1   Section: 3

37. Let $f(x) = \dfrac{e^x}{x^2 - 1}$. Find $f'(x)$.

    Answer: $\dfrac{e^x (x^2 - 2x - 1)}{(x^2 - 1)^2}$

    Difficulty: 1   Section: 3

38. Solve the equation $\ln |6 - x| = 2$.

    Answer: $x = 6 - e^2$ or $x = 6 + e^2$

    Difficulty: 2   Section: 3

39. Let $y = 2^\pi e^x$. Find $\dfrac{dy}{dx}$.

    Answer: $2^\pi e^x$

    Difficulty: 1   Section: 3

40. Let $f(x) = e^{-\sqrt{x}}$. Find $f'(x)$.

    Answer: $-\dfrac{e^{-\sqrt{x}}}{2\sqrt{x}}$

    Difficulty: 2   Section: 3

41. Let $y = \sqrt{e^x} + e^{-x}$. Find $\dfrac{dy}{dx}$.

    Answer: $\dfrac{1}{2}\sqrt{e^x} - e^{-x}$

    Difficulty: 2   Section: 3

42. Find $\dfrac{dy}{dx}$ and simplify if $y = \dfrac{\ln x}{e^x}$.

    Answer: $\dfrac{1 - x \ln x}{x e^x}$

    Difficulty: 2   Section: 3

43. Let $f(x) = x e^{\sin 3x}$. Find $f'(x)$.

    Answer: $e^{\sin 3x} (3x \cos 3x + 1)$

    Difficulty: 2   Section: 3

44. Let $f(x) = e^{x \ln x}$. Find $f'(x)$.

    Answer: $e^{x \ln x} (1 + \ln x)$

Difficulty: 2   Section: 3

45. Let $y = \dfrac{e^x - e^{-x}}{e^x + e^{-x}}$. Find $\dfrac{dy}{dx}$.

    Answer: $\dfrac{4}{(e^x + e^{-x})^2}$

    Difficulty: 2   Section: 3

46. Let $f(x) = e^x \sin x + e^x \cos x$. Find $f'(x)$.

    Answer: $2e^x \cos x$

    Difficulty: 2   Section: 3

47. Let $f(x) = \dfrac{1}{\sqrt{3e}} e^{\sin 3x}$. Find the slope of the tangent line to the graph of $f(x)$ at the point where $x = \dfrac{\pi}{18}$.

    Answer: $\dfrac{3}{2}$

    Difficulty: 2   Section: 3

48. Use implicit differentiation to find $\dfrac{dy}{dx}$ if $ye^x - \ln xy = 4$.

    Answer: $\dfrac{y - xy^2 e^x}{xye^x - x}$

    Difficulty: 2   Section: 3

49. Find $\displaystyle\int \dfrac{e^{4x}}{\sqrt{1 - e^{4x}}}\, dx$.

    Answer: $-\dfrac{1}{2}\sqrt{1 - e^{4x}} + C$

    Difficulty: 2   Section: 3

50. Find $\displaystyle\int \dfrac{e^{4x}}{1 + e^{4x}}\, dx$.

    Answer: $\dfrac{1}{4} \ln\left(1 + e^{4x}\right) + C$

    Difficulty: 2   Section: 3

51. Find $\displaystyle\int \dfrac{(e^x + 2)^2}{e^x}\, dx$.

    Answer: $e^x + 2x - e^{-x} + C$

    Difficulty: 2   Section: 3

52. Find $\int xe^{x^2}\, dx$.

   Answer: $\dfrac{1}{2}e^{x^2} + C$

   Difficulty: 2   Section: 3

53. Find $\int \dfrac{e^{\sqrt{x}}}{\sqrt{x}}\, dx$.

   Answer: $2e^{\sqrt{x}} + C$

   Difficulty: 1   Section: 3

54. Evaluate $\int_0^2 \left(e^x - 3x^2\right)\, dx$.

   Answer: $e^2 - 9$

   Difficulty: 1   Section: 3

55. Evaluate $\int_1^3 x\left(e^{x^2} - 1\right)\, dx$.

   Answer: $\dfrac{1}{2}\left(e^8 - 1\right)$

   Difficulty: 2   Section: 3

56. Evaluate $\int_1^e \dfrac{1}{x}\ln\, dx$.

   Answer: $\dfrac{1}{2}$

   Difficulty: 2   Section: 3

57. Find the area bounded by the curves $y = e^{x/2}$, the y-axis, the x-axis, and the line $x = 2$.

   Answer: $2(e - 1)$ square units

   Difficulty: 2   Section: 3

58. Find the volume of the solid generated by revolving about the x-axis, the region bounded by $y = e^{-x}$, the x-axis, and the lines $x = 0$ and $x = 3$.

   Answer: $\dfrac{\pi}{2}\left(1 - e^{-6}\right)$ cubic units.

   Difficulty: 2   Section: 3

59. Let $f(x) = 9^x$. Find $f'(x)$.

   Answer: $(\ln 9)(9^x)$

   Difficulty: 1   Section: 4

60. Let $f(x) = \cos(x \cdot 3^x)$. Find $f'(x)$.

   Answer: $-3^x \sin(x \cdot 3^x)(x \ln 3 + 1)$

   Difficulty: 2   Section: 4

61. Let $y = 3\ln(9^x)$. Find $\dfrac{dy}{dx}$.

   Answer: $3\ln 9$

   Difficulty: 1   Section: 4

62. Let $f(x) = 3^{2-x}2$. Find $f'(x)$.

   Answer: $-2(\ln 3)x\left(3^{2-x}2\right)$

   Difficulty: 2   Section: 4

63. Find $\dfrac{dy}{dx}$ for $y = 2^{\sin x}$.

   Answer: $(\ln 2)\left(2^{\sin x}\right)(\cos x)$

   Difficulty: 1   Section: 4

64. Find $\dfrac{dy}{dx}$ for $y = x^{\sin x}$.

   Answer: $\left(x^{\sin x}\right)\left(\dfrac{\sin x}{x} + (\cos x)(\ln x)\right)$

   Difficulty: 2   Section: 4

65. Find $\dfrac{dy}{dx}$ for $y = (\ln x)^{\ln x}$.

   Answer: $\left((\ln x)^{\ln x}\right)\left(\dfrac{1 + \ln(\ln x)}{x}\right)$

   Difficulty: 2   Section: 4

66. Find $\displaystyle\int 3^{2x}\,dx$.

   Answer: $\left(\dfrac{1}{2\ln 3}\right)3^{2x} + C$

   Difficulty: 1   Section: 4

67. Find $\displaystyle\int x^2 3^{x^3}\,dx$.

   Answer: $\dfrac{3^{x^3}}{3\ln 3} + C$

   Difficulty: 2   Section: 4

68. Find $\int 10^{1-x}\,dx$.

Answer: $-\dfrac{10^{1-x}}{\ln 10} + C$

Difficulty: 2   Section: 4

69. After 2 years at 9% compounded monthly what will the value of an initial investment of $7000, rounded to the nearest dollar?

Answer: $8375

Difficulty: 1   Section: 5

70. What is the value of an investment of $1 after 1 year at 100% annual interest compounded daily? (rounded to the nearest cent).

Answer: $2.71

Difficulty: 1   Section: 5

71. If an investment of $3000 is compounded continuously for 2 years at 10% what will it be worth? (round to the nearest dollar).

Answer: $3664

Difficulty: 1   Section: 5

72. The population of a certain town is approximated by the function $N(t) = 2500e^{0.05t}$, where t is measured in years after 1980. What will be the percent of increase in population from 1981 to 1990? (round to the nearest percent).

Answer: 57%

Difficulty: 2   Section: 5

73. Find the doubling time for a population given by the function $P(t) = 450{,}000e^{.02t}$.

Answer: $50 \ln 2 \approx 34.7$ years

Difficulty: 2   Section: 5

74. A bone is found to contain 30% of the carbon-14 that it contained when it was part of a living organism. How long ago did the organism die? (The half life of carbon-14 is 5568 years).

Answer: $\dfrac{5568\,[\ln(0.3)]}{\ln(0.50)} \approx 9670$ years ago

Difficulty: 2   Section: 5

75. The population P of a certain country is given by the function $P(t) = 450{,}000e^{kt}$, where t is the time in years from 1990 and k is constant. If the population was 495,000 in 1995, what will the population be in 2010 (answer to the nearest thousandth)?

Answer: 659,000

Difficulty: 2   Section: 5

76. Solve the differential equation $x\dfrac{dy}{dx} + y = x^3$.

   Answer: $y = \dfrac{x^3}{4} + \dfrac{C}{x}$

   Difficulty: 1   Section: 6

77. Solve the differential equation $x\dfrac{dy}{dx} + y = x^3$ if $y = 2$ when $x = 2$.

   Answer: $y = \dfrac{x^3}{4}$

   Difficulty: 1   Section: 6

78. Solve the differential equation $x\dfrac{dy}{dx} - y = x^3$.

   Answer: $y = \dfrac{x^3}{2} + Cx$

   Difficulty: 1   Section: 6

79. Solve the differential equation $x\dfrac{dy}{dx} - y = x^3$ if $y = 2$ when $x = 2$.

   Answer: $y = \dfrac{x^3}{2} - x$

   Difficulty: 1   Section: 6

80. Solve the differential equation $x\dfrac{dy}{dx} + 2y = \sin x$.

   Answer: $y = \dfrac{\sin x - x\cos x + C}{x^2}$

   Difficulty: 1   Section: 6

81. Solve the differential equation $x\dfrac{dy}{dx} + 2y = \sin x$ if $y = 0$ when $x = \pi$.

   Answer: $y = \dfrac{\sin x - x\cos x - \pi}{x^2}$

   Difficulty: 2   Section: 6

82. Solve the differential equation $x\dfrac{dy}{dx} - 2y = x^3$.

   Answer: $y = x^3 + Cx^2$

   Difficulty: 1   Section: 6

83. Solve the differential equation $x\dfrac{dy}{dx} - 2y = x^3$ if $y = 1$ when $x = 2$.

   Answer: $y = x^3 - \dfrac{7}{4}x^2$

   Difficulty: 1   Section: 6

84. Solve the differential equation $x\dfrac{dy}{dx} + 3y = x^3 + x$.

   Answer: $y = \dfrac{x^3}{6} + \dfrac{x}{4} + \dfrac{C}{x^3}$

   Difficulty: 1   Section: 6

85. Solve the differential equation $x\dfrac{dy}{dx} - 3y = x^3 + 2x$.

   Answer: $y = x^3 \ln x - x + Cx^3$

   Difficulty: 2   Section: 6

86. The figure below contains a slope field for the differential equation $y' = f(x,y)$. Sketch the solution to this differential equation that satisfies $y(0) = 0.5$ and approximate $y(3)$.

   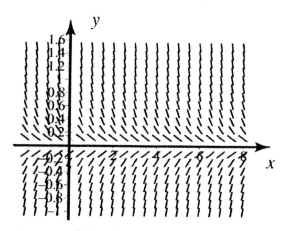

   Answer: $y(3) \approx 0$

   Difficulty: 1:   Section: 7

87. The figure below contains a slope field for the differential equation $y' = f(x,y)$. Sketch the solution to this differential equation that satisfies $y(0) = 0.2$ and approximate $y(2)$.

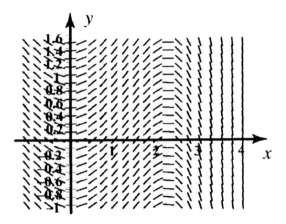

Answer: $y(2) \approx 1.5$

Difficulty: 1:   Section: 7

88. The figure below contains a slope field for the differential equation $y' = f(x,y)$. Sketch the solution to this differential equation that satisfies $y(0) = 4$ and approximate $y(8)$.

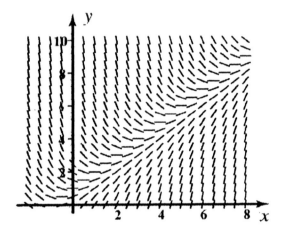

Answer: $y(8) \approx 8$

Difficulty: 1:   Section: 7

89. Use Euler's Method with $h = 0.2$ to approximate $y(1)$ where y is the solution of $y' = 3y$ that satisfies $y(0) = 1$.

Answer: 10.48576

Difficulty: 2:   Section 7

90. Use Euler's Method with $h = 0.2$ to approximate $y(2)$ where y is the solution of $y' = xy$ that satisfies $y(1) = 2$.

Answer: 6.838419

Difficulty: 2:   Section 7

91. Determine $\sin^{-1}(-1)$.

   Answer: $-\dfrac{\pi}{2}$

   Difficulty: 1   Section: 8

92. Determine $\cos^{-1}\left(-\dfrac{1}{2}\right)$.

   Answer: $\dfrac{2\pi}{3}$

   Difficulty: 1   Section: 8

93. Determine $\tan^{-1}\left(\tan\dfrac{3\pi}{4}\right)$.

   Answer: $-\dfrac{\pi}{4}$

   Difficulty: 1   Section: 8

94. Determine $\sec^{-1}\left(-\dfrac{2}{\sqrt{3}}\right)$.

   Answer: $\dfrac{5\pi}{6}$

   Difficulty: 2   Section: 8

95. Determine $\sin^{-1}\left(-\dfrac{\sqrt{3}}{2}\right) + \tan^{-1}(-1)$.

   Answer: $-\dfrac{7\pi}{12}$

   Difficulty: 2   Section: 8

96. Determine $\sin^{-1}(0) + \tan^{-1}\left(-\dfrac{1}{\sqrt{3}}\right)$.

   Answer: $-\dfrac{\pi}{6}$

   Difficulty: 2   Section: 8

97. Determine $\cos^{-1}\left(\sin\dfrac{\pi}{12}\right)$.

   Answer: $\dfrac{5\pi}{12}$

   Difficulty: 2   Section: 8

98. Determine $\tan\left[\cos^{-1}\left(\dfrac{3}{\sqrt{13}}\right)\right]$.

Answer: $\dfrac{2}{3}$

Difficulty: 2   Section: 8

99. Determine $\tan\left(\sin^{-1} x\right)$ in terms of x.

Answer: $\dfrac{x}{\sqrt{1-x^2}}$

Difficulty: 2   Section: 8

100. Determine $\sin\left(\tan^{-1} \dfrac{5}{2x}\right)$ in terms of x.

Answer: $\dfrac{5}{\sqrt{4x^2+25}}$

Difficulty: 2   Section: 8

101. Determine $\sin\left[\tan^{-1}\left(\dfrac{x-2}{4}\right)\right]$ in terms of x.

Answer: $\dfrac{x-2}{\sqrt{x^2-4x+20}}$

Difficulty: 2   Section: 8

102. Determine $\tan\left(\sin^{-1}\sqrt{1-x^2}\right)$ in terms of x.

Answer: $\dfrac{\sqrt{1-x^2}}{|x|}$

Difficulty: 2   Section: 8

103. Determine $\sin\left(\cos^{-1}\dfrac{\sqrt{9-4x^2}}{3}\right)$ in terms of x.

Answer: 2

Difficulty: 2   Section: 8

104. Let $f(x) = \tan^{-1}\left(\dfrac{x-3}{1+3x}\right)$. Find $f'(x)$ and simplify.

Answer: $\dfrac{1}{x^2+1}$

Difficulty: 2   Section: 8

105. Let $f(x) = x\tan^{-1}\left(\dfrac{x}{2}\right)$. Find $f'(x)$.

Answer: $\tan^{-1}\left(\dfrac{x}{2}\right) + \dfrac{2x}{x^2+4}$

Difficulty: 2   Section: 8

106. Let $y = \tan^{-1}\left(\sqrt{\ln(x^2)}\right)$. Find $\dfrac{dy}{dx}$.

   Answer: $\dfrac{1}{x\left[1 + \ln(x^2)\right]\sqrt{\ln(x^2)}}$

   Difficulty: 2   Section: 8

107. Find $\dfrac{dy}{dx}$ if $y = \tan^{-1}(\sin 2x)$.

   Answer: $\dfrac{2\cos 2x}{1 + \sin^2 2x}$

   Difficulty: 2   Section: 8

108. Let $f(x) = x^2 \sec^{-1}(e^x)$. Find $f'(x)$.

   Answer: $\dfrac{x^2}{\sqrt{e^{2x} - 1}} + 2x\sec^{-1}(e^x)$

   Difficulty: 2   Section: 8

109. Let $y = \sin^{-1}(1 - 2x)$. Find $\dfrac{dy}{dx}$.

   Answer: $-\dfrac{1}{\sqrt{x - x^2}}$

   Difficulty: 2   Section: 8

110. Let $y = \tan^{-1}(\tan^{-1} x)$. Find $\dfrac{dy}{dx}$.

   Answer: $\dfrac{1}{(1 + x^2)\left[1 + (\tan^{-1} x)^2\right]}$

   Difficulty: 2   Section: 8

111. Evaluate $\displaystyle\int_{3/4}^{3/2} \dfrac{2}{\sqrt{9 - 4x^2}}\, dx$.

   Answer: $\dfrac{\pi}{3}$

   Difficulty: 2   Section: 8

112. Evaluate $2\displaystyle\int_0^{1/\sqrt{2}} \dfrac{x}{\sqrt{1 - x^4}}\, dx$.

   Answer: $\dfrac{\pi}{6}$

   Difficulty: 2   Section: 8

113. Evaluate $\displaystyle\int_0^{\pi/2} \dfrac{\cos x}{1 + \sin^2 x}\, dx$.

Answer: $\dfrac{\pi}{4}$

Difficulty: 2   Section: 8

114. Find $\displaystyle\int \dfrac{1}{\sqrt{9-16x^2}}\,dx$.

    Answer: $\dfrac{1}{2}\sin^{-1}\left(\dfrac{4x}{3}\right) + C$

    Difficulty: 2   Section: 8

115. Find $\displaystyle\int \dfrac{e^{2x}}{1+e^{4x}}\,dx$.

    Answer: $\dfrac{1}{2}\tan^{-1}\left(e^{2x}\right) + C$

    Difficulty: 2   Section: 8

116. Find $\displaystyle\int \dfrac{3}{x\sqrt{x^4-1}}\,dx$.

    Answer: $\dfrac{3}{2}\sec^{-1}\left(x^2\right) + C$ or $-\dfrac{3}{2}\tan^{-1}\dfrac{1}{\sqrt{(x^4-1)}}$

    Difficulty: 1   Section: 8

117. Find $\displaystyle\int \dfrac{1}{\sqrt{\dfrac{25}{4} - (x-2)^2}}\,dx$.

    Answer: $\sin^{-1}\left(\dfrac{2x-4}{5}\right) + C$

    Difficulty: 1   Section: 8

118. Find $\displaystyle\int \dfrac{x^2}{7+x^6}\,dx$.

    Answer: $\dfrac{1}{3\sqrt{7}}\tan^{-1}\left(\dfrac{x^3}{\sqrt{7}}\right) + C$

    Difficulty: 1   Section: 8

119. Find $\displaystyle\int \dfrac{e^{2x}+e^{4x}}{\sqrt{1-e^{4x}}}\,dx$.

    Answer: $\dfrac{1}{2}\sin^{-1}\left(e^{2x}\right) - \dfrac{1}{2}\sqrt{1-e^{4x}} + C$

    Difficulty: 2   Section: 8

120. Find $\displaystyle\int \dfrac{1}{x\sqrt{1-(\ln x^2)}}\,dx$.

Answer: $\sin^{-1}(\ln x) + C$

Difficulty: 2   Section: 8

121. Let $f(x) = \cosh^2 3x$. Find $f'(x)$.

Answer: $6 \sinh 3x \cosh 3x = 3 \sinh 6x$

Difficulty: 1   Section: 9

122. Let $f(x) = \tanh(x^2 + 1)$. Find $f'(x)$.

Answer: $2x \operatorname{sech}^2(x^2 + 1)$

Difficulty: 1   Section: 9

123. Let $f(x) = 3e^{\sinh(3x^2)}$. Find $f'(x)$.

Answer: $18x \cosh(3x^2) e^{\sinh(3x^2)}$

Difficulty: 2   Section: 9

124. Let $f(x) = \dfrac{\sinh x}{1 + \cosh x}$. Find $f'(x)$.

Answer: $\dfrac{1}{1 + \cosh x}$

Difficulty: 2   Section: 9

125. Let $y = \ln|1 - \cosh 3x|$. Find $\dfrac{dy}{dx}$.

Answer: $\dfrac{-3 \sinh 3x}{1 - \cosh 3x}$

Difficulty: 2   Section: 9

126. Let $f(x) = \tanh(\ln|h|)$. Find $f'(x)$.

Answer: $\dfrac{1}{x}\left[\operatorname{sech}^2(\ln|x|)\right]$

Difficulty: 2   Section: 9

127. Find $\int x^2 \sinh(3x^3)\, dx$.

Answer: $\dfrac{1}{9} \cosh(3x^3) + C$

Difficulty: 1   Section: 9

128. Find $\int \tanh 2x\, dx$.

Answer: $\dfrac{1}{2} \ln(\cosh 2x) + C$

Difficulty: 1   Section: 9

129. Find $\int \cosh 2x \sqrt{\sinh 2x}\, dx$.

   Answer: $\dfrac{1}{3}(\sinh 2x)^{3/2} + C$

   Difficulty: 1   Section: 9

130. Find $\int x \cosh(x^2)\, dx$.

   Answer: $\dfrac{1}{2}\sinh(x^2) + C$

   Difficulty: 1   Section: 9

131. Find $\int \csc h 4x \coth 4x\, dx$.

   Answer: $-\dfrac{1}{4}\csc h 4x + C$

   Difficulty: 1   Section: 9

132. Find $\int \dfrac{\sinh(\ln x)}{3x}\, dx$.

   Answer: $\dfrac{1}{3}\cosh(\ln x) + C$

   Difficulty: 1   Section: 9

133. Let $f(x) = 4\sinh^{-1}(4x)$. Find $f'(x)$.

   Answer: $\dfrac{16}{\sqrt{1+16x^2}}$

   Difficulty: 1   Section: 9

134. Let $f(x) = \cosh^{-1}(2\cos x)$. Find $f'(x)$.

   Answer: $-\dfrac{2\sin x}{\sqrt{4\cos^2 x - 1}}$

   Difficulty: 2   Section: 9

135. Let $f(x) = \tanh^{-1}(\sin x)$. Find $f'(x)$.

   Answer: $\sec x$

   Difficulty: 2   Section: 9

136. Find $\int \dfrac{e^x}{\sqrt{e^{2x}-1}}\, dx$.

   Answer: $\cosh^{-1}(e^x) + C$

Difficulty: 2   Section: 9

137. Find $\int \dfrac{1}{\sqrt{25x^2+4}}\,dx$.

Answer: $\dfrac{1}{5}\sinh^{-1}\left(\dfrac{5x}{2}\right)+C$

Difficulty: 2   Section: 9

7 Techniques of Integration

1. Show that $\int \dfrac{1}{x^2 + 2x + 5}\,dx = \dfrac{1}{2}\tan^{-1}\left[\dfrac{1}{2}(x+1)\right] + C.$

   Answer: The derivative of the right hand side of the equation is equal to the function being integrated.

   Difficulty: 1   Section: 1

2. Show that $\int x \ln x\,dx = \dfrac{1}{2}x^2 \ln x - \dfrac{1}{4}x^2 + C.$

   Answer: The derivative of the right hand side of the equation is equal to the function being integrated.

   Difficulty: 1   Section: 1

3. Show that $\int e^x \cos x\,dx = \dfrac{1}{2}e^x(\sin x + \cos x) + C.$

   Answer: The derivative of the right hand side of the equation is equal to the function being integrated.

   Difficulty: 1   Section: 1

4. Show that $\int \cos^3 x \sin^4 x\,dx = \dfrac{1}{5}\sin^5 x - \dfrac{1}{7}\sin^7 x + C.$

   Answer: The derivative of the right hand side of the equation is equal to the function being integrated.

   Difficulty: 1   Section: 1

5. Show that $\int \dfrac{1}{\sqrt{4+x^2}}\,dx = \ln\left(x + \sqrt{4+x^2}\right) + C.$

   Answer: The derivative of the right hand side of the equation is equal to the function being integrated.

   Difficulty: 1   Section: 1

6. Find $\int \dfrac{1}{x^2 + 4x + 5}\,dx.$

   Answer: $\tan^{-1}(x+2) + C$

   Difficulty: 2   Section: 1

7. Find $\int \dfrac{1}{6 + 3x^2}\,dx.$

   Answer: $\dfrac{\sqrt{2}}{6}\tan^{-1}\left(\dfrac{x}{\sqrt{2}}\right) + C$

Difficulty: 1   Section: 1

8. Find $\int \dfrac{1}{\sqrt{x^2+8x+25}}\,dx$.

   Answer: $\sinh^{-1}\left(\dfrac{x+4}{3}\right)+C$

   Difficulty: 1   Section: 1

9. Find $\int \dfrac{3x+3}{\sqrt{x^2+2x-5}}\,dx$.

   Answer: $3\sqrt{x^2+2x-5}+C$

   Difficulty: 1   Section: 1

10. Find $\int \dfrac{2}{\sqrt{2x-x^2}}\,dx$.

    Answer: $2\sin^{-1}(x-1)+C$

    Difficulty: 1   Section: 1

11. Find $\int \dfrac{x-5}{x^2+9}\,dx$.

    Answer: $\dfrac{1}{2}\ln(x^2+9)-\dfrac{5}{3}\tan^{-1}\left(\dfrac{x}{3}\right)+C$

    Difficulty: 2   Section: 1

12. Find $\int \dfrac{2x-1}{x^2-6x+13}\,dx$.

    Answer: $\ln|x^2-6x+13|+\dfrac{5}{2}\tan^{-1}\left(\dfrac{x-3}{2}\right)+C$

    Difficulty: 2   Section: 1

13. Find $\int \dfrac{2x-1}{x^2+4x+8}\,dx$.

    Answer: $\ln|x^2+4x+8|-\dfrac{5}{2}\tan^{-1}\left(\dfrac{x+2}{2}\right)+C$

    Difficulty: 2   Section: 1

14. Find $\int \dfrac{1}{\sqrt{9+16x-4x^2}}\,dx$.

    Answer: $\dfrac{1}{2}\sin^{-1}\left(\dfrac{2x+4}{5}\right)+C$

    Difficulty: 2   Section: 1

15. Evaluate $\int_0^{2\sqrt{3}-2} \dfrac{1}{x^2+4x+8}\, dx$.

    Answer: $\dfrac{\pi}{24}$

    Difficulty: 1   Section: 1

16. Evaluate $\int_{3/4}^{3/2} \dfrac{1}{\sqrt{9-4x^2}}\, dx$.

    Answer: $\dfrac{\pi}{6}$

    Difficulty: 2   Section: 1

17. Evaluate $\int_0^{\sqrt{3}-1} \dfrac{1}{\sqrt{3-2x-x^2}}\, dx$.

    Answer: $\dfrac{\pi}{6}$

    Difficulty: 2   Section: 1

18. Find $\int x^2 e^{2x}\, dx$.

    Answer: $\dfrac{1}{2}e^{2x}\left(x^2 - x + \dfrac{1}{2}\right) + C$

    Difficulty: 1   Section: 2

19. Find $\int x e^{3x}\, dx$.

    Answer: $\dfrac{1}{3}xe^{3x} - \dfrac{1}{9}e^{3x} + C$

    Difficulty: 1   Section: 2

20. Find $\int \cos\sqrt{x}\, dx$.

    Answer: $2\sqrt{x}\sin\sqrt{x} + 2\cos\sqrt{x} + C$

    Difficulty: 2   Section: 2

21. Find $\int \sin^{-1} x\, dx$.

    Answer: $x\sin^{-1} x + \sqrt{1-x^2} + C$

    Difficulty: 2   Section: 2

22. Find $\int x\tan^{-1} x\, dx$.

Answer: $\dfrac{1}{2}\left[x^2 \tan^{-1} x + \tan^{-1} x - x\right] + C$

Difficulty: 2   Section: 2

23. Find $\displaystyle\int x^2 \tan^{-1} x \, dx$.

    Answer: $\dfrac{1}{3}x^3 \tan^{-1} x - \dfrac{1}{6}x^2 + \dfrac{1}{6}\ln\left|x^2 + 1\right| + C$

    Difficulty: 2   Section: 2

24. Find $\displaystyle\int x \ln x \, dx$.

    Answer: $\dfrac{1}{2}x^2 \left(\ln x - \dfrac{1}{2}\right) + C$

    Difficulty: 2   Section: 2

25. Find $\displaystyle\int \ln\left(x^2 + 1\right) dx$.

    Answer: $x \ln\left|x^2 + 1\right| - 2x + 2\tan^{-1} x + C$

    Difficulty: 2   Section: 2

26. Find $\displaystyle\int \ln\left(x^2 + 2\right) dx$.

    Answer: $x \ln\left(x^2 + 2\right) - 2x + 2\sqrt{2}\tan^{-1}\left(\dfrac{x}{2}\right) + C$

    Difficulty: 2   Section: 2

27. Find $\displaystyle\int (\ln x)^2 \, dx$.

    Answer: $x(\ln x)^2 - 2x \ln x + 2x + C$

    Difficulty: 2   Section: 2

28. Find $\displaystyle\int \sqrt{x} \ln x \, dx$.

    Answer: $\dfrac{2}{3}x^{3/2} \ln x - \dfrac{4}{9}x^{3/2} + C$

    Difficulty: 2   Section: 2

29. Evaluate $\displaystyle\int_1^4 \sqrt{x} \ln x \, dx$.

    Answer: $\dfrac{16}{3}\ln 4 + -\dfrac{28}{9}$

    Difficulty: 2   Section: 2

30. Evaluate $\int_1^e x \ln(x^2)\, dx$.

    Answer: $\dfrac{e^2 + 1}{2}$

    Difficulty: 2   Section: 2

31. Find $\int x \sin^{-1}\left(\dfrac{3}{x}\right) dx$ (for $x > 0$).

    Answer: $\dfrac{x^2}{2} \sin^{-1}\left(\dfrac{3}{x}\right) + \dfrac{3}{2}\sqrt{x^2 - 9} + C$

    Difficulty: 2   Section: 2

32. Find $\int x^3 \sqrt{1 - x^2}\, dx$.

    Answer: $\dfrac{1}{5}\left(1 - x^2\right)^{5/2} - \dfrac{1}{3}\left(1 - x^2\right)^{3/2} + C$

    Difficulty: 2   Section: 2

33. Find $\int \sin^5 7x\, dx$.

    Answer: $-\dfrac{1}{7}\left(\cos 7x - \dfrac{2\cos^3 7x}{3} + \dfrac{\cos^5 7x}{5}\right) + C$

    Difficulty: 2   Section: 3

34. Find $\int \cos^3 4x\, dx$.

    Answer: $\dfrac{1}{4} \sin 4x - \dfrac{\sin^3 4x}{12} + C$

    Difficulty: 2   Section: 3

35. Find $\int \cos^2 5x\, dx$.

    Answer: $\dfrac{x}{2} + \dfrac{\sin 10x}{20} + C$

    Difficulty: 2   Section: 3

36. Find $\int \sin^3\left(\dfrac{x}{2}\right) dx$.

    Answer: $-2\cos\left(\dfrac{x}{2}\right) + \dfrac{2}{3}\cos^3\left(\dfrac{x}{2}\right) + C$

    Difficulty: 2   Section: 3

37. Find $\int \sin x \left(1 + \cos^2 x\right) dx$.

Answer: $-\cos x - \dfrac{1}{3}\cos^3 x + C$

Difficulty: 1   Section: 3

38. Find $\displaystyle\int \cos^2 x \sin^4 x\, dx$.

Answer: $\dfrac{1}{8}\left[\dfrac{x}{2} - \dfrac{\sin 4x}{8} - \dfrac{\sin^3 2x}{6}\right] + C$

Difficulty: 2   Section: 3

39. Find $\displaystyle\int \cos^3 x \sin^4 x\, dx$.

Answer: $\dfrac{1}{5}\sin^5 x - \dfrac{1}{7}\sin^7 x + C$

Difficulty: 2   Section: 3

40. Find $\displaystyle\int \sin^3 x \cos^2 x\, dx$.

Answer: $-\dfrac{\cos^3 x}{3} + \dfrac{\cos^5 x}{5} + C$

Difficulty: 2   Section: 3

41. Find $\displaystyle\int \sin^2 x \cos^3 x\, dx$.

Answer: $\dfrac{1}{3}\sin^3 x - \dfrac{1}{5}\sin^5 x + C$

Difficulty: 2   Section: 3

42. Find $\displaystyle\int \dfrac{\cos^3 x}{\sqrt{\sin x}}\, dx$.

Answer: $2\sqrt{\sin x} - \dfrac{2}{5}\sqrt{\sin^5 x} + C$

Difficulty: 2   Section: 3

43. Evaluate $\displaystyle\int_0^{\pi/4} \cos^2 x \sin^4 x\, dx$.

Answer: $\dfrac{1}{16}\left(\dfrac{\pi}{4} - \dfrac{1}{3}\right)$

Difficulty: 2   Section: 3

44. Evaluate $\displaystyle\int_{\pi/3}^{\pi} \dfrac{\cos^2 x}{\csc x}\, dx$.

Answer: $\dfrac{3}{8}$

Difficulty: 2   Section: 3

45. Evaluate $\displaystyle\int_0^{\pi/2} \cos^3\left(\frac{x}{2}\right)\sin^4\left(\frac{x}{2}\right)\,dx$.

    Answer: $\sqrt{2}\left(\dfrac{1}{20} - \dfrac{1}{56}\right)$

    Difficulty: 2   Section: 3

46. Evaluate $\displaystyle\int_0^{\pi/40} \sin^2 5x \cos^2 5x\,dx$.

    Answer: $\dfrac{\pi - 2}{320}$

    Difficulty: 2   Section: 3

47. Find $\displaystyle\int \sin 3x \sin 2x\,dx$.

    Answer: $\dfrac{1}{2}\sin x - \dfrac{1}{10}\sin 5x + C$

    Difficulty: 1   Section: 3

48. Find $\displaystyle\int \cos x \cos 5x\,dx$.

    Answer: $\dfrac{1}{2}\left(\dfrac{\sin 6x}{6} + \dfrac{\sin 4x}{4}\right) + C$

    Difficulty: 1   Section: 3

49. Find $\displaystyle\int \tan^2 x \sec^2 x\,dx$.

    Answer: $\dfrac{1}{3}\tan^3 x + C$

    Difficulty: 1   Section: 3

50. Find $\displaystyle\int \tan^3 x \sec^3 x\,dx$.

    Answer: $\dfrac{\sec^5 x}{5} - \dfrac{\sec^3 x}{3} + C$

    Difficulty: 1   Section: 3

51. Find $\displaystyle\int \tan^6\left(\dfrac{x}{3}\right)\sec^2\left(\dfrac{x}{3}\right)\,dx$.

    Answer: $\dfrac{3\tan^7\left(\dfrac{x}{3}\right)}{7} + C$

    Difficulty: 1   Section: 3

52. Find $\int \tan^5 x \sec^3 x \, dx$.

   Answer: $\dfrac{\sec^7 x}{7} - \dfrac{2\sec^5 x}{5} + \dfrac{\sec^3 x}{3} + C$

   Difficulty: 2   Section: 3

53. Find $\int \tan^5 x \sec^4 x \, dx$.

   Answer: $\dfrac{\tan^8 x}{8} + \dfrac{\tan^6 x}{6} + C$

   Difficulty: 2   Section: 3

54. Find $\int \tan^3 x \, dx$.

   Answer: $\dfrac{\tan^2 x}{2} + \ln|\cos x| + C$

   Difficulty: 2   Section: 3

55. Find $\int \tan^5 3x \, dx$.

   Answer: $\dfrac{\tan^4 3x}{12} - \dfrac{\tan^2 3x}{6} - \dfrac{1}{3} \ln|\cos 3x| + C$

   Difficulty: 2   Section: 3

56. Find $\int \sec 5x \tan^3 5x \, dx$.

   Answer: $\dfrac{\sec^3 5x}{15} - \dfrac{\sec 5x}{5} + C$

   Difficulty: 2   Section: 3

57. Find $\int \dfrac{\cot \sqrt[3]{x}}{\sqrt[3]{x^2}} \, dx$.

   Answer: $3 \ln|\sin \sqrt[3]{x}| + C$

   Difficulty: 1   Section: 3

58. Find $\int (5 - \cot 3x)^2 \, dx$.

   Answer: $24x - \dfrac{10}{3} \ln|\sin 3x| - \dfrac{1}{3} \cot 3x + C$

   Difficulty: 2   Section: 3

59. Find $\int \dfrac{1}{4} \cot^4 \csc^2 x \, dx$.

Answer: $-\dfrac{1}{20}\cot^5 x + C$

Difficulty: 2   Section: 3

60. Find $\displaystyle\int (\csc x - 1)^2\,dx$.

    Answer: $-\cot x + 2\ln|\csc x + \cot x| + x + C$

    Difficulty: 2   Section: 3

61. Find $\displaystyle\int (\cot x + \csc x)\,dx$.

    Answer: $\ln|\sin x| - \ln|\csc x + \cot x| + C$

    Difficulty: 3   Section: 3

62. Find $\displaystyle\int \dfrac{x}{\sqrt{x^2 + 2x - 3}}\,dx$.

    Answer: $\sqrt{x^2 + 2x - 3} = \ln\left|x + 1\sqrt{x^2 + 2x - 3}\right| + C$

    Difficulty: 2   Section: 3

63. Find $\displaystyle\int \dfrac{1}{1 + \sqrt{x}}\,dx$.

    Answer: $2\left(\sqrt{x} - \ln|\sqrt{x} + 1|\right) + C$

    Difficulty: 2   Section: 3

64. Find $\displaystyle\int \dfrac{x^{1/2}}{4\left(x^{3/4} + 1\right)}\,dx$.

    Answer: $\dfrac{1}{3}x^{3/4}\dfrac{1}{3}\ln\left|x^{3/4} + 1\right| + C$

    Difficulty: 2   Section: 3

65. Find $\displaystyle\int \dfrac{1}{\sqrt{x} + \sqrt[3]{x}}\,dx$.

    Answer: $2x^{1/2} - 3x^{1/3} + 6x^{1/6} - 6\ln\left|x^{1/6} + 1\right| + C$

    Difficulty: 2   Section: 4

66. Find $\displaystyle\int \dfrac{\cot \sqrt[3]{x}}{\sqrt[3]{x^2}}\,dx$.

    Answer: $3\ln|\sin \sqrt{x}| + C$

    Difficulty: 1   Section: 4

67. Find $\displaystyle\int x\sqrt{x + 4}\,dx$.

Answer: $\dfrac{2}{5}(x+4)^{5/2} - \dfrac{8}{3}(x+4)^{3/2} + C$

Difficulty: 2   Section: 4

68. Find $\displaystyle\int \dfrac{x^3}{\sqrt[3]{x^2+4}}\, dx$.

Answer: $\dfrac{3}{10}\left(x^2+4\right)^{5/3} - 3\left(x^2+4\right)^{2/3} + C$

Difficulty: 2   Section: 4

69. Find $\displaystyle\int \sqrt{1-x^2}\, dx$.

Answer: $\dfrac{1}{2}\sin^{-1} x + \dfrac{1}{2}x\sqrt{1-x^2} + C$

Difficulty: 1   Section: 4

70. Find $\displaystyle\int \sqrt{9-4x^2}\, dx$.

Answer: $\dfrac{9}{4}\left[\sin^{-1}\left(\dfrac{2x}{3}\right) + \dfrac{2x\sqrt{9-4x^2}}{9}\right] + C$

Difficulty: 2   Section: 4

71. Find $\displaystyle\int \dfrac{x^2}{\sqrt{9-x^2}}\, dx$.

Answer: $\dfrac{9}{2}\sin^{-1}\left(\dfrac{x}{3}\right) - \dfrac{1}{2}x\sqrt{9-x^2} + C$

Difficulty: 2   Section: 4

72. Find $\displaystyle\int \dfrac{\sqrt{4-x^2}}{x^2}\, dx$.

Answer: $-\left[\dfrac{\sqrt{4-x^2}}{x} + \sin^{-1}\left(\dfrac{x}{2}\right)\right] + C$

Difficulty: 2   Section: 4

73. Find $\displaystyle\int \dfrac{1}{x^2\sqrt{4-x^2}}\, dx$.

Answer: $-\dfrac{\sqrt{4-x^2}}{4x} + C$

Difficulty: 2   Section: 4

74. Find $\displaystyle\int \dfrac{1}{\sqrt{9+4x^2}}\, dx$.

Answer: $\frac{1}{2}\ln\left|\sqrt{9-4x^2}+2x\right|+C$

Difficulty: 2   Section: 4

75. Find $\int \dfrac{\sqrt{x^2+16}}{x}\,dx$.

Answer: $\sqrt{x^2-16}-4\tan^{-1}\left(\dfrac{\sqrt{x^2-16}}{4}\right)+C$

Difficulty: 2   Section: 4

76. Find $\int \dfrac{x^3}{\sqrt{25-x^2}}\,dx$.

Answer: $\dfrac{1}{3}(25-x^2)^{3/2}-25(25-x^2)^{1/2}+C$

Difficulty: 2   Section: 4

77. Find $\int \dfrac{1}{\sqrt{9+16x-4x^2}}\,dx$.

Answer: $\dfrac{1}{2}\sin^{-1}\left(\dfrac{2x-4}{5}\right)+C$

Difficulty: 2   Section: 4

78. Find $\int \dfrac{1}{\sqrt{x^2+8x+25}}\,dx$.

Answer: $\ln\left|\sqrt{x^2+8x+25}+(x+4)\right|+C$

Difficulty: 2   Section: 4

79. Find $\int \dfrac{x}{\sqrt{x^2+2x-3}}\,dx$.

Answer: $\sqrt{x^2+2x-3}-\ln\left|x^2+2x-3\right|+C$

Difficulty: 2   Section: 4

80. Find $\int \dfrac{(4-x^2)^{3/2}}{x^6}\,dx$.

Answer: $-\dfrac{1}{20}\left(\dfrac{\sqrt{4-x^2}}{x}\right)^5+C$

Difficulty: 2   Section: 4

81. Find $\int_3^6 \dfrac{\sqrt{x^2-9}}{x}\,dx$.

Answer: $3\sqrt{3} - \pi$

Difficulty: 2 Section: 4

82. Use partial fractions to find $\int \dfrac{x+1}{x^2(x-1)}\,dx$.

Answer: $\dfrac{1}{x} + 2\ln|x-1| - 2\ln|x| + C$

Difficulty: 1 Section: 5

83. Use partial fractions to find $\int \dfrac{x+1}{x^3+x^2-6x}\,dx$.

Answer: $\dfrac{3}{10}\ln|x-2| - \dfrac{1}{6}\ln|x| - \dfrac{2}{15}\ln|x+3| + C$

Difficulty: 1 Section: 5

84. Use partial fractions to find $\int \dfrac{x+1}{x^2(x-1)(x^2+1)}\,dx$.

Answer: $-2\ln|x| + \dfrac{1}{x} + \ln|x-1| + \dfrac{1}{2}\ln|x^2+1| + C$

Difficulty: 2 Section: 5

85. Use partial fractions to evaluate $\displaystyle\int_0^4 \dfrac{1}{x^2+3x+2}\,dx$.

Answer: $\ln\dfrac{5}{3}$

Difficulty: 1 Section: 5

86. Use partial fractions to find $\int \dfrac{3}{x^3+x}\,dx$.

Answer: $3\ln|x| - \dfrac{3}{2}\ln|x^2+1| + C$

Difficulty: 1 Section: 5

87. Use partial fractions to find $\int \dfrac{3x^2+3x+2}{x^3+2x^2+x}\,dx$.

Answer: $2\ln|x| + \ln|x+1| + \dfrac{2}{x+1} + C$

Difficulty: 2 Section: 5

88. Use partial fractions to find $\int \dfrac{x^2-13x+30}{(x-1)(x-4)^2}\,dx$.

Answer: $2\ln|x-1| - \ln|x-4| + \dfrac{2}{x-4} + C$

Difficulty: 2   Section: 5

89. Use partial fractions to find $\displaystyle\int \frac{1}{x\,(x+1)^2}\, dx$.

    Answer: $\ln|x| - \ln|x-1| + \dfrac{1}{x+1} + C$

    Difficulty: 1   Section: 5

90. Use partial fractions to find $\displaystyle\int \frac{2x^2 - 12x + 4}{x^3 - 4x^2}\, dx$.

    Answer: $\dfrac{11}{4}\ln|x| + \dfrac{1}{x} - \dfrac{3}{4}\ln|x-4| + C$

    Difficulty: 2   Section: 5

91. Use partial fractions to find $\displaystyle\int \frac{x^3 - 4x - 1}{x\,(x-1)^3}\, dx$.

    Answer: $\ln|x| - \dfrac{3}{x-1} + \dfrac{2}{(x-1)^2} + C$

    Difficulty: 2   Section: 5

92. Use partial fractions to find $\displaystyle\int \frac{5x^2 + 30x + 43}{(x+3)^3}\, dx$.

    Answer: $5\ln|x+3| + \dfrac{1}{(x+3)^2} + C$

    Difficulty: 2   Section: 5

93. Use partial fractions to find $\displaystyle\int \frac{-2x}{(x+1)(x^2+1)}\, dx$.

    Answer: $\ln|x+1| - \dfrac{1}{2}\ln|x^2+1| - \tan^{-1} x + C$

    Difficulty: 2   Section: 5

94. Use partial fractions to find $\displaystyle\int \frac{5x^2 + 11x + 17}{(x^2+4)(x+5)}\, dx$.

    Answer: $\ln|x^2+4| + \dfrac{1}{2}\tan^{-1}\left(\dfrac{x}{2}\right) + 3\ln|x+5| + C$

    Difficulty: 2   Section: 5

95. Use partial fractions to find $\displaystyle\int \frac{5x^3 - 3x^2 + 2x - 1}{x^4 + x^2}\, dx$.

    Answer: $2\ln|x| + \dfrac{1}{x} + \dfrac{3}{2}\ln|x^2+1| - 2\tan^{-1} x + C$

    Difficulty: 2   Section: 5

96. Use partial fractions to find $\int \dfrac{x^2 - x - 21}{(x^2 + 4)(2x - 1)} \, dx$.

Answer: $\dfrac{3}{2} \ln|x^2 + 4| + \dfrac{1}{2} \tan^{-1}\left(\dfrac{x}{2}\right) - \dfrac{5}{2} \ln|2x - 1| + C$

Difficulty: 2   Section: 5

97. Use the table of integrals to evaluate $\int e^x \sin^{-1} e^x \, dx$.

Answer: $e^x \sin^{-2} e^x + \sqrt{1 - e^{2x}} + C$

Difficulty 1   Section 6

98. Use the table of integrals to evaluate $\int \dfrac{\sin x \cos x}{\sin x + 1} \, dx$.

Answer: $\sin x - \ln(\sin x + 1) + C$

Difficulty 2   Section 6

99. Find the value of c such that $\int_0^c e^{-2x} \, dx = \dfrac{1}{4}$.

Answer: $\dfrac{1}{2} \ln 2$

Difficulty 1   Section 6

100. Approximate the value of c that satisfies $\int_0^c e^{-x^2} \, dx = \dfrac{1}{4}$.

Answer: $0.255\,45$

Difficulty 2   Section 6

8 Indeterminate Forms and Improper Integrals

1. Find the limit if it exists: $\lim\limits_{x \to 0} \dfrac{(\sin x)^2}{\sin x^2}$.

   Answer: 1

   Difficulty: 1   Section: 1

2. Find the limit if it exists: $\lim\limits_{x \to \pi/2} \dfrac{\cos^2 x}{1 - \sin x}$.

   Answer: 2

   Difficulty: 1   Section: 1

3. Find the limit if it exists: $\lim\limits_{x \to 0} \dfrac{\sinh 3x}{x^3}$.

   Answer: limit does not exist

   Difficulty: 1   Section: 1

4. Find the limit if it exists: $\lim\limits_{x \to 0} \dfrac{\tan^{-1} x - x}{x^3}$.

   Answer: $-\dfrac{1}{3}$

   Difficulty: 1   Section: 1

5. Find the limit if it exists: $\lim\limits_{x \to 2} \dfrac{(x - 2) - \ln(x - 1)}{(x - 2)^2}$.

   Answer: $\dfrac{1}{2}$

   Difficulty: 1   Section: 1

6. Find the limit if it exists: $\lim\limits_{x \to 1} \dfrac{\tan \frac{\pi}{2} x}{x - 1}$.

   Answer: $-\infty$

   Difficulty: 1   Section: 1

7. Find the limit if it exists: $\lim\limits_{x \to 0} \dfrac{\tan^{-1} 2x}{\sin^{-1} x}$.

   Answer: 2

   Difficulty: 1   Section: 1

8. Find the limit if it exists: $\lim\limits_{x \to 0} \dfrac{e^{2x} - 1}{x}$.

Answer: 2

Difficulty: 1   Section: 1

9. Find the limit if it exists: $\lim\limits_{x \to \infty} \dfrac{x^3 - 4x^2 + 7x - 9}{x^2 + 7x - 5}$.

   Answer: ∞

   Difficulty: 1   Section: 2

10. Find the limit if it exists: $\lim\limits_{x \to \infty} \dfrac{4x^4 - 3x^3 + 2x^2 + 1009}{x^5 - 8x + 2,347}$.

    Answer: 0

    Difficulty: 1   Section: 2

11. Find the limit if it exists: $\lim\limits_{x \to \infty} \dfrac{x \sin x}{\sqrt{x}}$.

    Answer: limit does not exist

    Difficulty: 1   Section: 2

12. Find the limit if it exists: $\lim\limits_{x \to \infty} \dfrac{2^x + x^2}{3^x}$.

    Answer: 0

    Difficulty: 1   Section: 2

13. Find the limit if it exists: $\lim\limits_{x \to 0} x^2 \ln x$.

    Answer: 0

    Difficulty: 1   Section: 2

14. Find the limit if it exists: $\lim\limits_{x \to \infty} x \left(\tan^{-1} x - \dfrac{\pi}{2} \right)$.

    Answer: -1

    Difficulty: 1   Section: 2

15. Find the limit if it exists: $\lim\limits_{x \to \infty} x^2 e^{-2x}$.

    Answer: zero

    Difficulty: 1   Section: 2

16. Find the limit if it exists: $\lim\limits_{x \to \infty} \csc x \left[\ln (1 + \sin 2x) \right]$.

    Answer: 2

    Difficulty: 2   Section: 2

17. Find the limit if it exists: $\lim\limits_{x \to 0} (x - \ln x)$. (Hint: Write the function as a quotient with denominator $\dfrac{1}{x}$.)

Answer: ∞

Difficulty: 2   Section: 2

18. Find the limit if it exists: $\lim\limits_{x \to \infty} \left(\sqrt{x^2 + 3} - \sqrt{x^2 + 6x} \right)$.

Answer: -3

Difficulty: 1   Section: 2

19. Find the limit if it exists: $\lim\limits_{x \to 0^+} \dfrac{x^3}{\ln x}$.

Answer: e^3

Difficulty: 1   Section: 2

20. Find the limit if it exists: $\lim\limits_{x \to \infty} \dfrac{x^3}{\ln x}$.

Answer: e^3

Difficulty: 1   Section: 2

21. Find the limit if it exists: $\lim\limits_{t \to \infty} \left(\dfrac{t-1}{t+1} \right)^t$.

Answer: e^2

Difficulty: 1   Section: 2

22. Find the limit if it exists: $\lim\limits_{x \to 0} \left(\dfrac{1}{e^x - 1} - \dfrac{1}{x} \right)$.

Answer: $-\dfrac{1}{2}$

Difficulty: 1   Section: 2

23. Find the limit if it exists: $\lim\limits_{x \to \infty} (1 - 3x)^{1/x}$.

Answer: e^{-3}

Difficulty: 1   Section: 2

24. Find the limit if it exists: $\lim\limits_{x \to \infty} \left(\dfrac{x+2}{x-1} \right)^x$.

Answer: e^3

Difficulty: 1   Section: 2

25. Find the limit if it exists: $\lim_{x \to 0^+} x \csc x$.

    Answer: 1

    Difficulty: 1   Section: 2

26. Find the limit if it exists: $\lim_{x \to \infty} x \left(e^{2/x} - 1\right)$.

    Answer: 2

    Difficulty: 1   Section: 2

27. Find the limit if it exists: $\lim_{x \to 1} \left(\dfrac{\ln x}{(x-1)^2} - \dfrac{1}{x-1}\right)$.

    Answer: $-\dfrac{1}{2}$

    Difficulty: 1   Section: 2

28. Determine whether the integral converges or diverges and find the value if it converges: $\displaystyle\int_0^\infty \dfrac{1}{(x+1)^2}\,dx$.

    Answer: converges to 1

    Difficulty: 1   Section: 3

29. Determine whether the integral converges or diverges and find the value if it converges: $\displaystyle\int_{1/2}^\infty \dfrac{1}{1+4x^2}\,dx$.

    Answer: converges to $\dfrac{\pi}{8}$

    Difficulty: 1   Section: 3

30. Determine whether the integral converges or diverges and find the value if it converges: $\displaystyle\int_0^\infty \dfrac{1}{\sqrt{x+1}}\,dx$.

    Answer: diverges

    Difficulty: 1   Section: 3

31. Determine whether the integral converges or diverges and find the value if it converges: $\displaystyle\int_2^\infty \dfrac{2}{x^2-1}\,dx$.

    Answer: converges to $\ln 3$

    Difficulty: 2   Section: 3

32. Does the integral: $\displaystyle\int_{-\infty}^\infty \dfrac{1}{1+4x^2}\,dx$ converge or diverge? If it converges, find the value.

Answer: converges to $\dfrac{\pi}{2}$

Difficulty: 1   Section: 3

33. Does the integral: $\displaystyle\int_1^\infty \dfrac{1}{1+x^3}\,dx$ converge or diverge?

   Answer: converges by comparison the integral $\displaystyle\int_1^\infty \dfrac{1}{x^3}\,dx$

   Difficulty: 2   Section: 4

34. Does the integral: $\displaystyle\int_{-1}^1 \dfrac{1}{x^{1/3}}\,dx$ converge or diverge? If it converges, find the value.

   Answer: converges to zero

   Difficulty: 1   Section: 4

35. Does the integral: $\displaystyle\int_{3/2}^3 \dfrac{1}{\sqrt{9-x^2}}\,dx$ converge or diverge? If it converges, find the value.

   Answer: converges to $\dfrac{3\sqrt{3}}{2}$

   Difficulty: 1   Section: 4

36. Does the integral: $\displaystyle\int_0^2 \dfrac{x}{(x-1)^3}\,dx$ converge or diverge? If it converges, find the value.

   Answer: diverges

   Difficulty: 2   Section: 4

37. Does the integral: $\displaystyle\int_0^1 \dfrac{1}{\sqrt{1-x^2}}\,dx$ converge or diverge? If it converges, find the value.

   Answer: converges to $\dfrac{\pi}{2}$

   Difficulty: 1   Section: 4

38. Does the integral: $\displaystyle\int_{-\infty}^\infty \dfrac{1}{x^2}\,dx$ converge or diverge? If it converges, find the value.

   Answer: diverges

   Difficulty: 2   Section: 4

9 Infinite Series

1. Let $s_n = \tan^{-1} n$ for each n. Find $\lim_{n \to \infty} s_n$ if it exists.

    Answer: $\dfrac{\pi}{2}$

    Difficulty: 1   Section: 1

2. Let $s_n = \dfrac{3^n + n}{4^n}$ for each n. Find $\lim_{n \to \infty} s_n$ if it exists.

    Answer: 0

    Difficulty: 1   Section: 1

3. Let $s_n = \left(-\dfrac{1}{n}\right)^n$ for each n. Find $\lim_{n \to \infty} s_n$ if it exists.

    Answer: 0

    Difficulty: 2   Section: 1

4. Let $s_n = 2^{1/n}$ for each n. Find $\lim_{n \to \infty} s_n$ if it exists.

    Answer: 1

    Difficulty: 1   Section: 1

5. Let $s_n = \left(1 + \dfrac{3}{n}\right)$ for each n. Find $\lim_{n \to \infty} s_n$ if it exists.

    Answer: e^3

    Difficulty: 1   Section: 1

6. Let $s_n = \dfrac{n^3 + 3n - 2}{4n^3 + n^2 - 19n}$ for each n. Find $\lim_{n \to \infty} s_n$ if it exists.

    Answer: $\dfrac{1}{4}$

    Difficulty: 1   Section: 1

7. Let $s_n = \dfrac{n!}{x^3}$ for each n. Find $\lim_{n \to \infty} s_n$ if it exists.

    Answer: ∞

    Difficulty: 2   Section: 1

8. Let $s_n = \dfrac{|\cos n|}{\sqrt{n}}$ for each n. Find $\lim_{n \to \infty} s_n$ if it exists.

    Answer: 0

Difficulty: 2   Section: 1

9. Define $s_n = 1 + \cos n\pi$ for all n. Find $\lim_{n\to\infty} s_n$ if it exists.

   Answer: the limit does not exist

   Difficulty: 2   Section: 1

10. Define $s_n = (1.001)^n$ for each n. Find $\lim_{n\to\infty} s_n$ if it exists.

    Answer: ∞

    Difficulty: 1   Section: 1

11. Define $s_n = (-.999)^n$ for each n. Find $\lim_{n\to\infty} s_n$ if it exists.

    Answer: 0

    Difficulty: 1   Section: 1

12. Define $s_n = \sqrt[n]{3x}$ for each n. Find $\lim_{n\to\infty} s_n$ if it exists.

    Answer: 1

    Difficulty: 2   Section: 1

13. Define $s_n = \dfrac{3^n}{n!}$ for each n. Show that the sequence is eventually decreasing and bounded from below.

    Answer: $\dfrac{s_{n+1}}{x_n} = \dfrac{3}{n+1} \leq 1$ if $n \geq 2$. Also, $s_n \geq 0$ for all n.

    Difficulty: 2   Section: 1

14. Define $s_n = \dfrac{x+1}{n+2}$ for each n. Show that the sequence is eventually decreasing and bounded from below.

    Answer: Let $f(x) = \dfrac{x+1}{x+2}$, then $f'(x) = (x+2)^{-2} \geq 0$ for $x \geq 1$ and $\dfrac{n+1}{n+2} \leq 1$ for all n.

    Difficulty: 2   Section: 1

15. Define $s_n = \dfrac{n+2}{n+1}$ for each n. Show that the sequence is eventually decreasing and bounded from below.

    Answer: Let $f(x) = \dfrac{x+2}{x+1}$, then $f'(x) = -(x+1)^{-2} \leq 0$ for $x \geq 1$. Also, $\dfrac{n+1}{n+1} \geq 1$ for all n.

    Difficulty: 2   Section: 1

16. Define $s_n = \dfrac{n}{2^n}$ for each n. Show that the sequence is eventually decreasing and bounded from below.

Answer: $\dfrac{s_n}{s_{n+1}} = \dfrac{n+1}{2n} \leq 1$ for $n \geq 2$ and $\dfrac{n}{2^n} \geq 0$ for all n.

Difficulty: 2   Section: 1

17. Define $s_n = 1 + \dfrac{(-1)^n}{n}$ for each n. Show that the sequence is monotone.

    Answer: For n even, $s_n \geq 1$ and for n odd, $s_n \leq 1$.

    Difficulty: 2   Section: 1

18. Define $s_n = \dfrac{2^n}{n}$ for each n. Show that the sequence is increasing but not convergent.

    Answer: $\dfrac{s_{n+1}}{s_n} = \dfrac{x+1}{2x} \leq$ for $n \geq 2$. But the sequence is unbounded, hence it doesn't converge

    Difficulty: 2   Section: 1

19. Find the sum of the infinite series $\sum\limits_{n=1}^{\infty} \dfrac{2^n}{3^n}$.

    Answer: 2

    Difficulty: 1   Section: 2

20. Find the sum of the infinite series $\sum\limits_{n=1}^{\infty} \dfrac{2^{n+1}}{3^n}$.

    Answer: 4

    Difficulty: 1   Section: 2

21. Find the sum of the infinite series $\sum\limits_{n=0}^{\infty} \dfrac{3^n}{5^n}$.

    Answer: $\dfrac{5}{2}$

    Difficulty: 1   Section: 2

22. Find the sum of the infinite series $\sum\limits_{n=0}^{\infty} (0.6)^n$.

    Answer: 2.5

    Difficulty: 1   Section: 2

23. Find the sum of the infinite series $\sum\limits_{n=0}^{\infty} (0.75)^n$.

    Answer: 4

    Difficulty: 1   Section: 2

24. Use partial fractions to find a formula for s_n, the nth partial sum for $\sum\limits_{n=1}^{\infty} \dfrac{1}{n^2 + 3n + 2}$. Does

the series converge?

Answer: $s_n = \dfrac{1}{2} - \dfrac{1}{n+2}$. The series converges to $\dfrac{1}{2}$.

Difficulty: 2   Section: 2

25. Use partial fractions to find a formula for s_n, the nth partial sum for $\displaystyle\sum_{n=3}^{\infty} \dfrac{1}{n^2 - 3n + 2}$. Does the series converge?

Answer: $s_n = 1 - \dfrac{1}{n-1}$. The series converges to 1.

Difficulty: 2   Section: 2

26. Write the repeating decimal $0.\overline{7}$ (in other words, $0.777\ldots$) as a geometric series and write the sum a fraction.

Answer: $0.\overline{7} = \displaystyle\sum_{n=1}^{\infty} 7(0.1)^n = \dfrac{7}{9}$

Difficulty: 2   Section: 2

27. Write the repeating decimal $0.\overline{5}$ (in other words, $0.555\ldots$) as a geometric series and write the sum a fraction.

Answer: $0.\overline{5} = \displaystyle\sum_{n=1}^{\infty} 5(0.1)^n = \dfrac{5}{9}$

Difficulty: 2   Section: 2

28. Write the repeating decimal $0.\overline{15}$ (in other words, $0.151515\ldots$) as a geometric series and write the sum as a fraction.

Answer: $0.\overline{15} = \displaystyle\sum_{n=1}^{\infty} 15(0.1)^n = \dfrac{5}{33}$

Difficulty: 2   Section: 2

29. Write the repeating decimal $0.\overline{51}$ (in other words, $0.515151\ldots$) as a geometric series and write the sum as a fraction.

Answer: $0.\overline{51} = \displaystyle\sum_{n=1}^{\infty} 51(0.1)^n = \dfrac{17}{33}$

Difficulty: 2   Section: 2

30. Does the series $\displaystyle\sum_{n=1}^{\infty} \dfrac{n}{8n+3}$ converge? If so, find the sum.

Answer: The series diverges because $\displaystyle\lim_{n\to\infty} \dfrac{n}{8n+3} \neq 0$.

Difficulty: 1   Section: 2

31. Does the series $\displaystyle\sum_{n=1}^{\infty} \left(1 + \dfrac{1}{n}\right)^n$ converge? If so, find the sum.

Answer: The series diverges because $\lim_{n\to\infty} \left(1 + \frac{1}{n}\right)^n \neq 0$.

Difficulty: 1   Section: 2

32. Does the series $\sum_{n=1}^{\infty} \frac{5}{n}$ converge? If so, find the sum.

Answer: The series diverges, it is a multiple of the harmonic series.

Difficulty: 1   Section: 2

33. Does the series $\sum_{n=1}^{\infty} \left(\frac{5}{2} - \frac{1}{2^n}\right)$ converge? If so, find the sum.

Answer: The series diverges because $\sum_{n=1}^{\infty} \frac{5}{n}$ diverges and $\sum_{n=1}^{\infty} \frac{1}{2^n}$ converges.

Difficulty: 1   Section: 2

34. Does the series $\sum_{n=1}^{\infty} \left(\frac{5}{2^n} + \frac{2}{5^n}\right)$ converge? If so, find the sum.

Answer: The series converges to $\frac{11}{2}$, it is the sum of two geometric series.

Difficulty: 1   Section: 2

35. Does the series $\sum_{n=1}^{\infty} \left(\sin n + \frac{1}{3^n}\right)$ converge? If so, find the sum.

Answer: The series diverges since $\lim_{n\to\infty} \left(\sin n + \frac{1}{3^n}\right) \neq 0$.

Difficulty: 1   Section: 2

36. Find the sum of the series $\sum_{n=3}^{\infty} 2^{-n}$.

Answer: $\frac{1}{4}$

Difficulty: 1   Section: 2

37. Find the sum of the series $\sum_{n=3}^{\infty} 4^{-n}$.

Answer: $\frac{1}{48}$

Difficulty: 1   Section: 2

38. Find the sum of the series $\sum_{n=1}^{\infty} (-1)^n 2^{-n}$.

Answer: This is a geometric series with $r = -\frac{1}{2}$, the sum is $-\frac{1}{3}$

Difficulty: 1   Section: 2

39. Find the sum of the series $\sum_{n=2}^{\infty} (-1)^n 3^{-n}$.

    Answer: This is a geometric series with $r = -\dfrac{1}{3}$, the sum is $\dfrac{1}{12}$.

    Difficulty: 1   Section: 2

40. Does the series $\sum_{n=1}^{\infty} \dfrac{1}{\sqrt[n]{n}}$ converge? If so, find the sum.

    Answer: The series diverges because $\lim\limits_{n \to \infty} \dfrac{1}{\sqrt[n]{n}} \neq 0$.

    Difficulty: 2   Section: 2

41. Use the integral test to test the series $\sum_{n=1}^{\infty} \dfrac{1}{1+4n^2}$ for convergence. If the series converges, find an upper bound for the sum.

    Answer: The series converges and the sum is less than or equal to $\dfrac{1}{5} + \dfrac{1}{2}\left(\dfrac{\pi}{2} - \tan^{-1} 2\right)$.

    Difficulty: 1   Section: 3

42. Use the integral test to test the series $\sum_{n=1}^{\infty} (n+1)^{-1/2}$ for convergence. If the series converges, find an upper bound for the sum.

    Answer: The series diverges.

    Difficulty: 1   Section: 3

43. Use the integral test to test the series $\sum_{n=1}^{\infty} \dfrac{n^2}{1+n^3}$ for convergence. If the series converges, find an upper bound for the sum.

    Answer: The series diverges

    Difficulty: 1   Section: 3

44. Use the integral test to test the series $\sum_{n=1}^{\infty} ne^{-2n}$ for convergence. If the series converges, find an upper bound for the sum.

    Answer: The series converges; an upper bound for the sum is $\dfrac{3}{2}e^{-1}$.

    Difficulty: 1   Section: 3

45. Does the series $\sum_{n=1}^{\infty} \dfrac{1}{n^{1.5}}$ converge? Give a reason for your conclusion.

    Answer: The series converges; it is a p-series with $p = 1.5 > 1$.

Difficulty: 1   Section: 3

46. Does the series $\sum_{n=1}^{\infty} \dfrac{1}{n^{0.9}}$ converge? Give a reason for your conclusion.

    Answer: The series diverges; it is a p-series with $p = 0.9 < 1$.

    Difficulty: 1   Section: 3

47. Can the integral test be used for the series $\sum_{n=2}^{\infty} \dfrac{n}{\ln n}$?

    Answer: No, $f(x) = \dfrac{x}{\ln x}$ is not a decreasing function.

    Difficulty: 2   Section: 3

48. Does the series $\sum_{n=1}^{\infty} \dfrac{1}{4n^2 - 1}$ converge? Try the integral test.

    Answer: yes

    Difficulty: 2   Section: 3

49. Does the series $\sum_{n=1}^{\infty} n^{-1.1}$ converge? Give a reason for your conclusion.

    Answer: Yes; it is a p-series with $p = 1.1 > 1$.

    Difficulty: 1   Section: 3

50. Use the integral test to show that $\sum_{n=1}^{\infty} n^{-4}$ converges and find an upper bound for the sum.

    Answer: $\int_{1}^{\infty} x^{-4}\, dx = \dfrac{1}{3}$, an upper bound is $\dfrac{7}{12}$

    Difficulty: 1   Section: 3

51. Does the series $\sum_{n=1}^{\infty} 4n^{-2}$ converge? Give reasons.

    Answer: Yes, it is a multiple of a convergent p-series, $p = 2$.

    Difficulty: 1   Section: 3

52. Does the series $\sum_{n=1}^{\infty} \left[2^{-n} + n^{-2}\right]$ converge? Give reasons.

    Answer: Yes, it is the sum of a convergent geometric series and a convergent p-series.

    Difficulty: 2   Section: 3

53. Does the series $\sum_{n=1}^{\infty} \dfrac{7}{3n^2 + 5}$ converge?

    Answer: Yes, compare with the p-series with $p = 2$.

Difficulty: 1   Section: 4

54. Does the series $\sum_{n=1}^{\infty} \dfrac{7}{3n+5}$ converge?

Answer: No, compare with the harmonic series

Difficulty: 1   Section: 4

55. Does the series $\sum_{n=1}^{\infty} \dfrac{5}{2^n + n^2}$ converge? Give reasons.

Answer: Yes, compare with the geometric series with $r = \dfrac{1}{2}$.

Difficulty: 2   Section: 4

56. Does the series $\sum_{n=1}^{\infty} \dfrac{5n}{2^n}$ converge?

Answer: Yes, use either the ratio or root test

Difficulty: 1   Section: 4

57. Does the series $\sum_{n=1}^{\infty} \dfrac{\left(1+\dfrac{1}{n}\right)^n}{3^n}$ converge? Give reasons.

Answer: Yes, use the root test

Difficulty: 2   Section: 4

58. Does the series $\sum_{n=1}^{\infty} \dfrac{2n-7}{3n^2 + 5n - 1}$ converge?

Answer: No, compare with the harmonic series.

Difficulty: 1   Section: 4

59. Does the series $\sum_{n=1}^{\infty} \dfrac{(n+1)^2}{n!}$ converge?

Answer: Yes, use the ratio test.

Difficulty: 1   Section: 4

60. Does the series $\sum_{n=1}^{\infty} \dfrac{(n+1)!}{2n!}$ converge?

Answer: No, the limit of the nth term is not zero

Difficulty: 2   Section: 4

61. Does the series $\sum_{n=1}^{\infty} \dfrac{2^n}{n^3 + 1}$ converge?

Answer: No, use the ratio test

Difficulty: 1   Section: 4

62. Does the series $\sum_{n=1}^{\infty} \dfrac{n^n}{n!}$ converge?

Answer: No, use the ratio test

Difficulty: 1   Section: 4

63. Does the series $\sum_{n=1}^{\infty} 4n^{-2}$ converge? Give reasons.

Answer: Yes, it is a multiple of a convergent p-series, $p = 2$.

Difficulty: 1   Section: 4

64. Does the series $\sum_{n=1}^{\infty} \dfrac{n^2 - 7n + 5}{n^n}$ converge?

Answer: Yes, use the ratio or root test.

Difficulty: 2   Section: 4

65. Does the series $\sum_{n=1}^{\infty} e^{1/n}$ converge?

Answer: No, the limit of the n-th term is not zero.

Difficulty: 1   Section: 4

66. Does the series $\sum_{n=1}^{\infty} \left(\dfrac{2n}{3n+1}\right)^n$ converge?

Answer: Yes, use the root test

Difficulty: 1   Section: 4

67. Does the series $\sum_{n=1}^{\infty} \dfrac{\cos^2 n}{n^2}$ converge? Give reasons.

Answer: Yes, $\dfrac{\cos^2 n}{n^2} \leq \dfrac{1}{n^2}$ for all n.

Difficulty: 1   Section: 4

68. Does the series $\sum_{n=1}^{\infty} \dfrac{2^n + 1}{2^{n+1}}$ converge?

Answer: No, the limit of the nth term is not 0.

Difficulty: 1   Section: 4

69. Does the series $\sum_{n=1}^{\infty} \pi^{-n} e^n$ converge?

Answer: Yes, it is a geometric series with $r = \dfrac{e}{\pi} < 1$.

Difficulty: 1   Section: 4

70. Does the series $\sum\limits_{n=1}^{\infty} \dfrac{1}{\sqrt{n^2+1}}$ converge?

    Answer: No, compare with the harmonic series

    Difficulty: 1   Section: 4

71. Does the series $\sum\limits_{n=1}^{\infty} (-1)^n \dfrac{2^n - 1}{n+1}$ converge?

    Answer: No, the limit of the nth term is not zero.

    Difficulty: 1   Section: 5

72. Does the series $\sum\limits_{n=1}^{\infty} (-1)^n \dfrac{n}{n^2+1}$ converge?

    Answer: Yes, by the alternating series test

    Difficulty: 1   Section: 5

73. Does the series $\sum\limits_{n=1}^{\infty} (-1)^n n^{-1/3}$ converge? Give reasons.

    Answer: Yes, by the alternating series test

    Difficulty: 1   Section: 5

74. Does the series $\sum\limits_{n=1}^{\infty} \dfrac{\cos n\pi}{n^2}$ converge?

    Answer: Yes, $\cos n\pi = (-1)^n$, so the alternating series test applies.

    Difficulty: 1   Section: 5

75. Does the series $\sum\limits_{n=1}^{\infty} (-1)^n \left(\dfrac{3}{2}\right)^n$ converge?

    Answer: No, the limit of the nth term is not zero.

    Difficulty: 1   Section: 5

76. How many terms of the series $\sum\limits_{n=1}^{\infty} \dfrac{1}{2n}$ must be summed to estimate the sum of the series with error less than .001?

    Answer: 500

    Difficulty: 2   Section: 5

77. How many terms of the series $\sum\limits_{n=1}^{\infty} (-1)^n n^{-1/2}$ must be summed to estimate the sum of the

series with error less than .01?

Answer: 10,000

Difficulty: 2   Section: 5

78. How many terms of the series $\sum_{n=1}^{\infty} (-1)^n \frac{1}{n!}$ must be summed to estimate the sum of the series with error less than .001?

Answer: 4

Difficulty: 2   Section: 5

79. Estimate the sum of the series $\sum_{n=1}^{\infty} (-1)^n \frac{1}{n^2}$ with error less than 0.1.

Answer: $-\frac{31}{36}$

Difficulty: 2   Section: 5

80. Estimate the sum of the series $\sum_{n=1}^{\infty} (-1)^{n+1} \frac{1}{2n+1}$ with error less than 0.1.

Answer: $\frac{1}{3} - \frac{1}{5} + \frac{1}{7} - \frac{1}{9}$

Difficulty: 2   Section: 5

81. Classify the series $\sum_{n=1}^{\infty} (-1)^n \frac{n}{3n+7}$ as absolutely convergent, conditionally convergent, or divergent.

Answer: Divergent, the limit of the nth term is not zero.

Difficulty: 1   Section: 5

82. Classify the series $\sum_{n=1}^{\infty} \frac{(-1)^n}{3n^2 + 5}$ as absolutely convergent, conditionally convergent, or divergent.

Answer: Absolutely convergent, compare the series of absolute values to a p-series with $p = 2$.

Difficulty: 2   Section: 5

83. Classify the series $\sum_{n=1}^{\infty} (-1)^{n+1} \left(\frac{3}{5}\right)^n$ as absolutely convergent, conditionally convergent, or divergent.

Answer: Absolutely convergent, the series of absolute values is a geometric series with $r = \frac{3}{5}$.

Difficulty: 1   Section: 5

84. Classify the series $\sum_{n=1}^{\infty} \frac{\cos n\pi}{n}$ as absolutely convergent, conditionally convergent, or divergent.

Answer: Conditionally convergent by the alternating series test, the series of absolute values is the harmonic series which diverges.

Difficulty: 1   Section: 5

85. Classify the series $\sum_{n=1}^{\infty} (-1)^{n+1} \dfrac{1}{5n^2+1}$ as absolutely convergent, conditionally convergent, or divergent.

Answer: Absolutely convergent, compare the series of absolute values to a p-series with $p = 2$.

Difficulty: 2   Section: 5

86. Classify the series $\sum_{n=1}^{\infty} (-1)^{2n+1} \dfrac{1}{\sqrt{n+5}}$ as absolutely convergent, conditionally convergent, or divergent.

Answer: Divergent, all terms are negative; compare to a p-series with $p = \dfrac{1}{2}$.

Difficulty: 1   Section: 5

87. Classify the series $\sum_{n=1}^{\infty} (-1)^{n+3} (0.6)^n$ as absolutely convergent, conditionally convergent, or divergent.

Answer: Absolutely convergent, the series of absolute values is a geometric series with $r = 0.6$.

Difficulty: 1   Section: 5

88. Classify the series $\sum_{n=1}^{\infty} (-1)^{n+2}$ as absolutely convergent, conditionally convergent, or divergent.

Answer: Divergent, the limit of the nth term is not zero.

Difficulty: 1   Section: 5

89. Classify the series $\sum_{n=1}^{\infty} (-1)^n \sin n$ as absolutely convergent, conditionally convergent, or divergent.

Answer: Divergent, the limit of the nth term is not zero.

Difficulty: 2   Section: 5

90. Classify the series $\sum_{n=1}^{\infty} (-1)^{n-1}$ as absolutely convergent, conditionally convergent, or divergent.

Answer: Divergent, the limit of the nth term is not zero.0

Difficulty: 2   Section: 5

91. Find the convergence set of $\sum_{n=1}^{\infty} \dfrac{x^n}{n^2}$.

Answer: $-1 \leq x \leq 1$.

Difficulty: 1   Section: 6

92. Find the convergence set of $\sum_{n=1}^{\infty} \frac{x^n}{n!}$.

    Answer: x can be any real number.

    Difficulty: 1   Section: 6

93. Find the convergence set of $\sum_{n=1}^{\infty} \frac{x^n}{3x+1}$.

    Answer: $-1 \leq x < 1$

    Difficulty: 1   Section: 6

94. Find the convergence set of $\sum_{n=1}^{\infty} \frac{2^n x^n}{n}$.

    Answer: $-\frac{1}{2} \leq x < \frac{1}{2}$

    Difficulty: 1   Section: 6

95. Find the convergence set of $\sum_{n=1}^{\infty} \frac{x^n}{3^n}$.

    Answer: $-3 < x < 3$

    Difficulty: 1   Section: 6

96. Find the convergence set of $\sum_{n=1}^{\infty} \frac{x^n}{\sqrt{n}}$.

    Answer: $-1 \leq x < 1$, the convergence at -1 is conditional.

    Difficulty: 1   Section: 6

97. Find the convergence set of $\sum_{n=1}^{\infty} \frac{x^n}{x^3}$.

    Answer: $-1 \leq x \leq 1$

    Difficulty: 1   Section: 6

98. Find the convergence set of $\sum_{n=1}^{\infty} \frac{(-1)^n x^n}{2^n n}$.

    Answer: $-2 < x \leq 2$

    Difficulty: 1   Section: 6

99. Find the convergence set of $\sum_{n=1}^{\infty} \frac{2^n x^{2n}}{n!}$.

Answer: all real numbers

Difficulty: 1   Section: 6

100. Find the convergence set of $\sum_{n=1}^{\infty} \frac{(x-2)^n}{3^n}$

Answer: $-1 < x < 5$

Difficulty: 1   Section: 6

101. Find the convergence set of $\sum_{n=1}^{\infty} \frac{8^{3n} x^{2n}}{n!}$.

Answer: all real numbers

Difficulty: 1   Section: 6

102. Find the convergence set of $\sum_{n=1}^{\infty} \frac{x^{2n}}{2^n}$.

Answer: $-\sqrt{2} < x < \sqrt{2}$

Difficulty: 1   Section: 6

103. Find the convergence set of $\sum_{n=0}^{\infty} \frac{3^{n-1}}{2^n} x^n$.

Answer: $-\frac{2}{3} < x < \frac{2}{3}$

Difficulty: 1   Section: 6

104. Find the convergence set of $\sum_{n=1}^{\infty} \frac{n}{4^{n+1}} (x-1)^n$.

Answer: $-3 < x < 5$

Difficulty: 2   Section: 6

105. Find the convergence set of $\sum_{n=1}^{\infty} \frac{(x+3)^n}{2^n n}$.

Answer: $-5 \leq x < -1$

Difficulty: 2   Section: 6

106. Find the convergence set of $\sum_{n=1}^{\infty} \frac{(-1)^n (x-2)^{2n-1}}{(2n-1)!}$.

Answer: all real numbers

Difficulty: 2   Section: 6

107. Find the convergence set of $\sum_{n=1}^{\infty} \frac{n(x-1)^n}{(3n-1) 3^n}$

Answer: $-2 < x < 4$

Difficulty: 2   Section: 6

108. Find the convergence set of $\sum_{n=1}^{\infty} \dfrac{(x+3)^{2n}}{\sqrt{3n}}$.

Answer: $-4 < x < -2$

Difficulty: 2   Section: 6

109. Find the convergence set of $\sum_{n=1}^{\infty} \dfrac{x^n}{n^2}$.

Answer: $-1 \leq x \leq 1$

Difficulty: 2   Section: 6

110. Find the convergence set of $\sum_{n=1}^{\infty} \dfrac{(x+4)^n}{3^n n^2}$.

Answer: $-7 \leq x \leq -1$

Difficulty: 2   Section: 6

111. Find the convergence set of $\sum_{n=1}^{\infty} \dfrac{2^n (x+1)^n}{n^2}$.

Answer: $-\dfrac{3}{2} \leq x \leq -\dfrac{1}{2}$

Difficulty: 2   Section: 6

112. Write a power series for $f(x) = \dfrac{x}{(1-x)^2}$. Give the convergence set.

Answer: $\sum_{n=1}^{\infty} n\, x^n$, $-1 < x < 1$

Difficulty: 2   Section: 7

113. Write a power series for $f(x) = \dfrac{x}{(1+x)^2}$. Give the convergence set.

Answer: $\sum_{n=1}^{\infty} (-1)^n n\, x^n$, $-1 < x < 1$

Difficulty: 1   Section: 7

114. Write a power series for $f(x) = \dfrac{x^2}{1-x}$. Give the convergence set.

Answer: $\sum_{n=1}^{\infty} x^{n+2}$, $-1 < x < 1$

Difficulty: 1   Section: 7

115. Write a power series for $f(x) = \dfrac{1}{2+3x}$. Give the convergence set.

Answer: $\displaystyle\sum_{n=1}^{\infty} \dfrac{3^n(-1)^n}{2^{n+1}} x^n,\ -\dfrac{2}{3} < x < \dfrac{2}{3}$

Difficulty: 2   Section: 7

116. Find the power series for $f(x) = \dfrac{x}{3+2x}$. Give the convergence set.

Answer: $\displaystyle\sum_{n=1}^{\infty} \dfrac{(-1)^n 2^n x^{n+1}}{3^{n+1}} x^n,\ -\dfrac{3}{2} < x < \dfrac{3}{2}$

Difficulty: 2   Section: 7

117. Find the power series for $f(x) = \dfrac{x^2}{1-3x}$. Give the convergence set.

Answer: $\displaystyle\sum_{n=1}^{\infty} 3^n x^{n+2},\ -\dfrac{1}{3} < x < \dfrac{1}{3}$

Difficulty: 2   Section: 7

118. Find the power series for $f(x) = \tan^{-1}\left(\dfrac{x}{2}\right)$. Give the convergence set.

Answer: $\displaystyle\sum_{n=1}^{\infty} \dfrac{(-1)^n x^{2n+1}}{(2n+1)\, 2^{2n+1}},\ -2 \le x \le 2$

Difficulty: 1   Section: 7

119. Find the power series for $f(x) = \tan^{-1}(3x)$. Give the convergence set.

Answer: $\displaystyle\sum_{n=1}^{\infty} \dfrac{(-1)^n - 3^{2n+1} x^{2n+1}}{2n+1},\ -\dfrac{1}{3} < x < \dfrac{1}{3}$

Difficulty: 1   Section: 7

120. Find the power series for $f(x) = \ln(1+3x)$. Give the convergence set.

Answer: $\displaystyle\sum_{n=1}^{\infty} \dfrac{(-1)^{n-1} 3^n x^n}{n},\ -\dfrac{1}{3} < x < \dfrac{1}{3}$

Difficulty: 1   Section: 7

121. Find the power series for $f(x) = x \ln(1-2x)$. Give the convergence set.

Answer: $\displaystyle\sum_{n=1}^{\infty} (-1) \dfrac{2^n x^{n+1}}{n},\ -\dfrac{1}{2} \le x < \dfrac{1}{2}$

Difficulty: 1   Section: 7

122. Find the power series for $f(x) = x^2 \tan^{-1}\left(\dfrac{x}{2}\right)$. Give the convergence set.

Answer: $\displaystyle\sum_{n=1}^{\infty} \dfrac{(-1)^n x^{2n+3}}{(2n+1)\, 2^{2n+1}},\ -2 \le x \le 2$

Difficulty: 1   Section: 7

123. Find the power series for $f(x) = \dfrac{1}{1-x^2}$. Give the convergence set.

Answer: $\displaystyle\sum_{n=1}^{\infty} x^{2n}, \ -1 < x < 1$

Difficulty: 1   Section: 7

124. Find the Taylor series for $f(x) = x^4 - x^3 + 4x$ in powers of $x+2$.

Answer: $16 - 40(x+2) + 30(x+2)^2 - 9(x+2)^3 + (x+2)^4$

Difficulty: 2   Section: 8

125. Find the Maclaurin series for $f(x) = \dfrac{1}{3+x}$.

Answer: $\displaystyle\sum_{n=0}^{\infty} (-1)^n \dfrac{x^n}{3^{n+1}}$

Difficulty: 1   Section: 8

126. Find the Maclaurin series for $f(x) = \cos x^2$.

Answer: $\displaystyle\sum_{n=0}^{\infty} \dfrac{(-1)^n x^{4n}}{(2n)!}$

Difficulty: 1   Section: 8

127. Write the Taylor series for $f(x) = \ln x^2$ in powers of $(x-1)$.

Answer: $\displaystyle\sum_{n=0}^{\infty} \dfrac{2(-1)^n (x-1)^{n+1}}{n+1}$

Difficulty: 2   Section: 8

128. Write the Taylor series for $f(x) = e^{-x}$ in powers of $(x-1)$.

Answer: $\displaystyle\sum_{n=0}^{\infty} \dfrac{e^{-1}(-1)^n (x-1)^n}{n!}$

Difficulty: 2   Section: 8

129. Write the terms through x^4 in the Maclaurin series for $f(x) = \sec 2x$ about $c = 0$.

Answer: $1 + 2x^2 + \dfrac{10}{3}x^4$

Difficulty: 1   Section: 8

130. Write the Taylor series in $(x-8)$ through the term in $(x-8)^2$ for $f(x) = x^{2/3}$ about $c = 8$.

Answer: $4 + \dfrac{1}{3}(x-8) = \dfrac{1}{144}(x-8)^2$

Difficulty: 1   Section: 8

131. Write the Taylor series in $(x-1)$ for $f(x) = x^3 - 3x^2 + 5x - 2$.

    Answer: $1 + 2(x-1) + (x-1)^3$

    Difficulty: 1   Section: 8

132. Write the Taylor series for $f(x) = \dfrac{1}{x}$ in powers of $(x-2)$

    Answer: $\displaystyle\sum_{n=0}^{\infty} \dfrac{(-1)^n (x-2)^n}{2^{n+1}}$

    Difficulty: 2   Section: 8

133. Write the Taylor series for $f(x) = x\cos x$.

    Answer: $\displaystyle\sum_{n=0}^{\infty} (-1)^n \dfrac{x^{2n+1}}{(2n)!}$

    Difficulty: 1   Section: 8

134. Write the first two nonzero terms of the Maclaurin series for $f(x) = \tan 2x$.

    Answer: $2x + \dfrac{8}{3}x^3$

    Difficulty: 1   Section: 8

135. Write the first two nonzero terms of the Maclaurin series for $f(x) = \tanh 2x$

    Answer: $2x - \dfrac{8}{3}x^3$

    Difficulty: 1   Section: 8

136. Write the first four terms of the binomial series for $f(x) = (1-x)^{1/2}$.

    Answer: $1 - \dfrac{1}{2}x - \dfrac{1}{8}x^2 - \dfrac{1}{16}x^3$

    Difficulty: 1   Section: 8

137. Write the first four terms of the binomial series for $f(x) = (1+x)^{1/3}$.

    Answer: $1 + \dfrac{1}{2}x - \dfrac{1}{9}x^2 + \dfrac{1}{81}x^3$

    Difficulty: 1   Section: 8

138. Write the first four terms of the binomial series for $f(x) = (1-x)^{-1/2}$.

    Answer: $1 + \dfrac{1}{2}x - \dfrac{3}{8}x^2 + \dfrac{5}{16}x^3$

    Difficulty: 1   Section: 8

139. Write the first four terms of the binomial series for $f(x) = (1+x)^{-3}$.

Answer: $1 - 3x + 6x^2 - 10x^3$

Difficulty: 1   Section: 8

140. Write the first four terms of the binomial series for $f(x) = (4+x)^{1/2}$.

Answer: $2 + \dfrac{x}{4} - \dfrac{1}{64}x^2 + \dfrac{1}{512}x^3$

Difficulty: 2   Section: 8

141. Write the Taylor polynomial of order 3 based at 1 for $f(x) = e^x$.

Answer: $e\left(1 + (x-1) + \dfrac{(x-1)^2}{2} + \dfrac{(x-1)^3}{6}\right)$

Difficulty: 1   Section: 9

142. Write the Taylor polynomial of order 3 based at 2 for $f(x) = e^x$.

Answer: $e^2\left(1 + (x-2) + \dfrac{(x-2)^2}{2} + \dfrac{(x-2)^3}{6}\right)$

Difficulty: 1   Section: 9

143. Write the Taylor polynomial of order 3 based at -1 for $f(x) = e^x$.

Answer: $e^{-1}\left(1 + (x11) + \dfrac{(x+1)^2}{2} + \dfrac{(x+1)^3}{3}\right)$

Difficulty: 1   Section: 9

144. Write the Taylor polynomial of order 3 based at -2 for $f(x) = e^x$.

Answer: $e^{-2}\left(1 + (x+2) + \dfrac{(x+2)^2}{2} + \dfrac{(x+2)^3}{2}\right)$

Difficulty: 1   Section: 9

145. Write the Maclaurin polynomial of order 4 for $f(x) = \sec 2x$.

Answer: $1 + 2x^2 + \dfrac{10}{3}x^4$

Difficulty: 1   Section: 9

146. Write the Maclaurin polynomial of order 4 for $f(x) = \sec 3x$.

Answer: $1 + \dfrac{9}{2}x^2 + \dfrac{135}{8}x^4$

Difficulty: 1   Section: 9

147. Write the Taylor polynomial of order 3 based at 1 for $f(x) = x^3 - 3x^2 + 5x - 2$.

　　Answer: $1 + 2(x-1) + (x-1)^3$

　　Difficulty: 1　　Section: 9

148. Write the Taylor polynomial of order 3 based at 2 for $f(x) = x^3 - 3x^2 + 5x - 2$.

　　Answer: $4 + 5(x-2) + 3(x-2)^2 + (x-2)^3$

　　Difficulty: 1　　Section: 9

149. Write the Maclaurin polynomial of order 3 for $\tan^{-1}\frac{x}{3}$.

　　Answer: $\frac{x}{3} - \frac{1}{81}x^3$

　　Difficulty: 1　　Section: 9

150. Write the Taylor polynomial of order 2 based at 8 for $f(x) = x^{2/3}$.

　　Answer: $4 + \frac{1}{3}(x-8) - \frac{1}{144}(x-8)^2$

　　Difficulty: 1　　Section: 9

151. Find a polynomial approximation $f(x) = e^x$ accurate within 0.01 for $-\frac{1}{2} \le x \le \frac{1}{2}$.

　　Answer: $1 + x + \frac{x^2}{2} + \frac{x^3}{6}$

　　Difficulty: 2　　Section: 9

152. Find a polynomial approximation $f(x) = \sin x$ accurate within 0.01 for $-1 \le x \le 1$.

　　Answer: $x - \frac{x^3}{6}$

　　Difficulty: 2　　Section: 9

153. Find a polynomial approximation $f(x) = \cos x$ accurate within 0.01 for $-1 \le x \le 1$.

　　Answer: $1 - \frac{x^2}{2} + \frac{x^4}{24}$

　　Difficulty: 2　　Section: 9

154. Determine the order of the Maclaurin polynomial required to approximate e^x correct to 2 decimal places for $|x| \le \frac{1}{2}$.

　　Answer: order 4 using 1.7 as an upper bound for $e^{1/2}$

　　Difficulty: 2　　Section: 9

155. Determine the order of the Maclaurin polynomial required to approximate e^x correct to 3

decimal places for $|x| \leq \dfrac{1}{2}$.

Answer: order 5 using 1.7 as an upper bound for $e^{1/2}$

Difficulty: 2   Section: 9

156. Determine the order of the Maclaurin polynomial required to approximate e^x correct to 4 decimal places for $|x| \leq \dfrac{1}{2}$.

Answer: order 6 using 1.7 as an upper bound for $e^{1/2}$

Difficulty: 2   Section: 9

157. Find the third order of Maclaurin polynomial for $(1+x)^{1/2}$ and bound the error $R_3(x)$ for $0 \leq x \leq \dfrac{1}{2}$.

Answer: $1 + \dfrac{1}{2}x - \dfrac{1}{8}x^2 + \dfrac{1}{16}x^3$, $|R_3(x)| \leq .019$

Difficulty: 2   Section: 9

158. Find the fourth order of Maclaurin polynomial for $(1+x)^{1/2}$ and bound the error $R_4(x)$ for $0 \leq x \leq \dfrac{1}{2}$.

Answer: $1 + \dfrac{1}{2}x - \dfrac{1}{8}x^2 + \dfrac{1}{16}x^3 - \dfrac{5}{128}x^4$, $|R_4(x)| \leq 1.008$

Difficulty: 2   Section: 9

159. Find $\lim\limits_{x \to 0} \dfrac{\ln(1+x) - x + \dfrac{x^2}{2}}{x^3}$.

Answer: $\dfrac{1}{3}$

Difficulty: 2   Section: 9

160. Find $\lim\limits_{x \to 0} \dfrac{\tan^{-1} x - x + \dfrac{x^3}{3}}{2x^5}$.

Answer: $\dfrac{1}{10}$

Difficulty: 2   Section: 9

161. Find $\lim\limits_{x \to 0} \dfrac{2x^5 - 7x^6}{\tan x - x - \dfrac{x^3}{3}}$.

Answer: 15

Difficulty: 2   Section: 9

162. Find $\lim\limits_{x\to 0} \dfrac{e^{-2x} - 1 + x^2}{3x^4 + 5x^7 - x^9}$.

Answer: $\dfrac{1}{6}$

Difficulty: 2   Section: 9

10 Conics and Polar Coordinates

1. Find the equation of the parabola with focus $(2,0)$ and directrix $x = -2$.

   Answer: $y^2 = 8x$

   Difficulty: 1   Section: 1

2. Sketch the parabola $y = 16x$ showing vertex, focus, and directrix on the graph.

   Answer:

   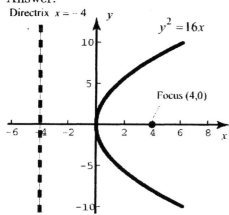

   Difficulty: 1   Section: 1

3. Sketch the parabola $x = \dfrac{1}{8}y^2$ showing vertex, focus, and directrix on the graph.

   Answer:

   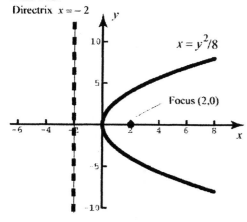

   Difficulty: 1   Section: 1

4. Sketch the parabola $x^2 = 8y$ showing vertex, focus, and directrix on the graph.

   Answer:

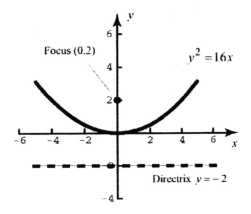

Difficulty: 1   Section: 1

5. Find the vertices on the major axis of the ellipse $3x^2 + 2y^2 = 16$.

   Answer: $(0, 2\sqrt{2})$ and $(0, -2\sqrt{2})$

   Difficulty: 1   Section: 2

6. Find the foci of the ellipse $4y^2 + x^2 = 36$.

   Answer: $(-\sqrt{27}, 0)$ and $(\sqrt{27}, 0)$

   Difficulty: 1   Section: 2

7. Sketch the ellipse $2x^2 + 4y^2 = 8$ showing center, vertices, and foci on the graph.

   Answer:

   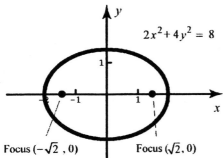

   Difficulty: 1   Section: 2

8. Sketch the ellipse $9x^2 + 4y^2 = 36$ showing center, vertices, and foci on the graph.

   Answer:

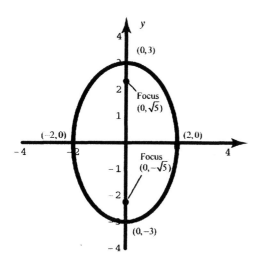

Difficulty: 1   Section: 2

9. Find the foci and vertices of the hyperbola $9x^2 - 4y^2 = 36$.

   Answer: foci $\left(\sqrt{13}, 0\right)$, $\left(-\sqrt{13}, 0\right)$ and vertices $(-2, 0)$, $(2, 0)$

   Difficulty: 1   Section: 2

10. Find the foci and vertices of the hyperbola $9x^2 - 16y^2 = -144$.

    Answer: foci $(0, 5)$, $(0, -5)$ and vertices $(0, 3)$, $(0, -3)$

    Difficulty: 1   Section: 2

11. Find the equation of the hyperbola with foci at $(5, 0)$, $(-5, 0)$ and vertices at $(4, 0)$, $(-4, 0)$.

    Answer: $\dfrac{x^2}{16} - \dfrac{y^2}{9} = 1$

    Difficulty: 1   Section: 2

12. Find the equation of the hyperbola with foci at $(0, 6)$, $(0, -6)$ and vertices at $(0, 2)$, $(0, -2)$.

    Answer: $8y^2 - x^2 = 32$

    Difficulty: 1   Section: 2

13. Find the equation of the hyperbola with foci at $(0, 2)$, $(0, -2)$ and an asymptote $y = 2x$.

    Answer: $\dfrac{5y^2}{16} - \dfrac{5y^2}{4} = 1$

    Difficulty: 1   Section: 2

14. Sketch the hyperbola $\dfrac{x^2}{25} - \dfrac{y^2}{4} = 1$ showing vertices, foci, and asymptotes on the graph.

    Answer: Foci are at $\left(\pm\sqrt{29}, 0\right)$.

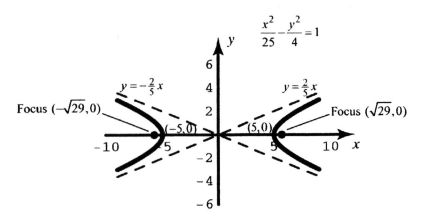

Difficulty: 2   Section: 2

15. Sketch the hyperbola $4y^2 - x^2 = 1$ showing vertices, foci, and asymptotes on the graph.

Answer: Foci are at $\left(0, \pm\sqrt{\dfrac{5}{2}}\right)$.

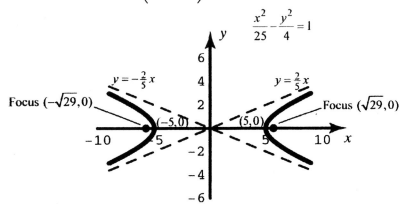

Difficulty: 2   Section: 2

16. Find the equation of the hyperbola with foci at $(5,0)$ and $(-5,0)$ and a fixed difference of distances equal to 6.

Answer: $16x^2 - 9y^2 = 144$

Difficulty: 1   Section: 2

17. Find the equation of the hyperbola with foci at $(0,8)$ and $(0,-8)$ and a fixed difference of distances equal to 10.

Answer: $\dfrac{y^2}{25} - \dfrac{x^2}{39} = 1$

Difficulty: 1   Section: 2

18. Find the equation of the ellipse with foci at $(-3,0)$ and $(3,0)$ and a fixed sum of distances equal to 12.

Answer: $\dfrac{x^2}{36} + \dfrac{y^2}{27} = 1$

Difficulty: 1   Section: 2

19. Find the equation of the ellipse with foci at $(0, 1)$ and $(0, -1)$ and a fixed sum of distances equal to 5.

    Answer: $\dfrac{4x^2}{21} + \dfrac{4y^2}{25} = 1$

    Difficulty: 1   Section: 2

20. Find the slope of the tangent line to the curve $x^2 - 4y^2 = 9$ at the point $(5, 2)$.

    Answer: $\dfrac{5}{8}$

    Difficulty: 1   Section: 2

21. Find the equation of the tangent line to the curve $9x^2 + 4y^2 = 72$ at the point $(-2, 3)$.

    Answer: $y - 3 = \dfrac{3}{2}(x + 2)$

    Difficulty: 2   Section: 2

22. Find the equation of the tangent line to the curve $4x^2 + 9y^2 = 40$ at the point $(1, 2)$.

    Answer: $y - 2 = -\dfrac{2}{9}(x - 1)$

    Difficulty: 2   Section: 2

23. Find the equation of the tangent line to the curve $x^2 - y^2 = 16$ at the point $(5, 3)$.

    Answer: $y - 3 = \dfrac{5}{3}(x - 5)$

    Difficulty: 2   Section: 2

24. Find the total area enclosed by the ellipse $\dfrac{x^2}{25} + \dfrac{y^2}{9} = 1$.

    Answer: 15π square units

    Difficulty: 2   Section: 2

25. Find the equation of the parabola with focus $(2, 9)$ and vertex $(2, 6)$.

    Answer: $(x - 2)^2 = 12(y - 6)$

    Difficulty: 1   Section: 3

26. Find the equation of the parabola with focus $(4, 6)$ and vertex $(1, 6)$.

    Answer: $(y - 6)^2 = 12(x - 1)$

    Difficulty: 1   Section: 3

27. Find the equation of the parabola with focus $(3,-1)$ and vertex $(-2,-1)$.

    Answer: $(y+1)^2 = 20(x+2)$

    Difficulty: 1   Section: 3

28. Find the equation of the parabola with focus $(-2,3)$ and directrix $y=5$.

    Answer: $(x+2)^2 = -4(y-4)$

    Difficulty: 1   Section: 3

29. Find the equation of the parabola with focus $(-3,2)$ and directrix $x=5$.

    Answer: $(y-2)^2 = -16(x-1)$

    Difficulty: 1   Section: 3

30. Find the vertex, focus, and directrix of the parabola $y^2 - 2y + 9 = 8x$.

    Answer: $(1,1)$, $(3,1)$, $x=-1$

    Difficulty: 2   Section: 3

31. Find the vertex, focus, and directrix of the parabola $x^2 + 2x + 8y = 15$.

    Answer: $(-1,2)$, $(-1,0)$, $y=4$

    Difficulty: 2   Section: 3

32. Find the vertex, focus, and directrix of the parabola $x^2 - 2x + 4y = 11$.

    Answer: $(1,3)$, $(1,2)$, $y=4$

    Difficulty: 2   Section: 3

33. Sketch the parabola $y^2 - 2y - 31 = 8x$ showing vertex, focus, and directrix on the graph.

    Answer:

    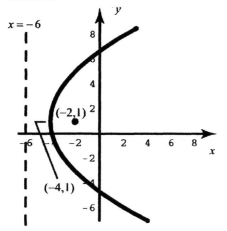

    Difficulty: 2   Section: 3

E168   Chapter 10   Exam Questions   Instructor's Resource Manual

34. Sketch the parabola $y^2 - 4y + 12 = 8x$ showing vertex, focus, and directrix on the graph.

    Answer:

    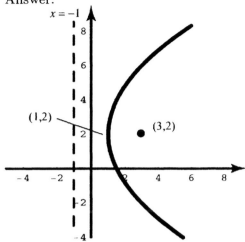

    Difficulty: 2   Section: 3

35. Find the center of the ellipse $4x^2 + y^2 + 16x - 4y = 80$.

    Answer: $(-2, 2)$

    Difficulty: 1   Section: 3

36. Find the foci of the ellipse $25x^2 + 4y^2 - 50x + 16y = 59$.

    Answer: $\left(1, -2 + \sqrt{21}\right)$ and $\left(1, -2 - \sqrt{21}\right)$

    Difficulty: 2   Section: 3

37. Find the foci of the ellipse $3x^2 - 6x + 2y^2 + 4y = 19$.

    Answer: $(1, 1)$ and $(1, -3)$

    Difficulty: 2   Section: 3

38. Find the equation of the ellipse with vertices at $(2, 4)$, $(2, -6)$, $(5, -1)$, and $(-1, -1)$.

    Answer: $\dfrac{(x-2)^2}{9} + \dfrac{(y+1)^2}{25} = 1$

    Difficulty: 2   Section: 3

39. Find the equation of the ellipse with vertices at $(3, 2)$, $(-7, 2)$, $(-2, 12)$, and $(-2, -8)$.

    Answer: $\dfrac{(x+2)^2}{25} + \dfrac{(y-2)^2}{100} = 1$

    Difficulty: 2   Section: 3

40. Find the equation of the ellipse with foci at $(-4, 2)$ and $(2, 2)$ and a fixed sum if distances equal to 12.

Answer: $\dfrac{(x+1)^2}{36} + \dfrac{(y-2)^2}{27} = 1$

Difficulty: 2   Section: 3

41. Find the equation of the ellipse with foci at $(4,-1)$ and $(-4,-1)$ and a fixed sum of distances equal to 10.

    Answer: $\dfrac{x^2}{25} + \dfrac{(y+1)^2}{9} = 1$

    Difficulty: 2   Section: 3

42. Find the equation of the ellipse with foci at $(1,1)$ and $(-3,1)$ and a fixed sum of distances equal to $2\sqrt{5}$.

    Answer: $\dfrac{(x+1)^2}{5} + (y-1)^2 = 1$

    Difficulty: 2   Section: 3

43. Sketch the ellipse $x^2 + 4y^2 + 6x - 16y + 9 = 0$ showing center, vertices, and foci on the graph.

    Answer:

    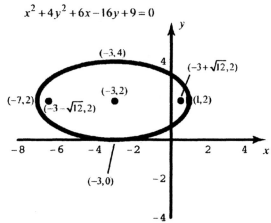

    Difficulty: 2   Section: 3

44. Sketch the ellipse $4x^2 + y^2 + 16x - 4y - 80 = 0$ showing center, vertices, and foci on the graph.

    Answer:

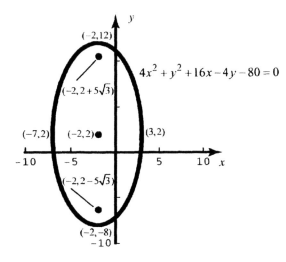

Difficulty: 2   Section: 3

45. Sketch the ellipse $25x^2 + 9y^2 - 200x + 90y + 400 = 0$ showing center, vertices, and foci on the graph.

Answer:

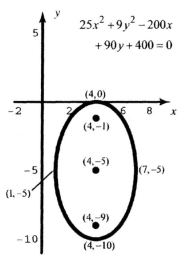

Difficulty: 2   Section: 3

46. Sketch the parabola $x^2 + 2x + 8y = 15$ showing vertex, focus, and directrix on the graph.

Answer:

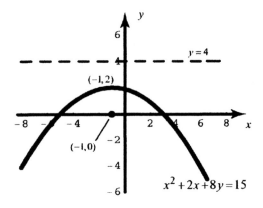

Difficulty: 2   Section: 3

47. Find the foci and vertices of the hyperbola $4x^2 - y^2 + 16x - 4y = 88$.

    Answer: foci$(-2 + 5\sqrt{5}, -2)$, $(-2 - 5\sqrt{5}, -2)$ and vertices $(-7, -2)$, $(3, -2)$

    Difficulty: 2   Section: 3

48. Find the foci and vertices of the hyperbola $25x^2 - 9y^2 - 50x - 18y = -241$.

    Answer: foci$(1, -1 + \sqrt{34})$, $(-1, -1 - \sqrt{34})$ and vertices $(1, 4)$, $(1, -6)$

    Difficulty: 2   Section: 3

49. Find the equation of the hyperbola with foci at $(5, -2)$, $(-5, 2)$ and vertices at $(4, -2)$, $(-4, -2)$.

    Answer: $9x^2 - 16(y + 2)^2 = 144$

    Difficulty: 1   Section: 3

50. Find the equation of the hyperbola with vertices at $(1, 2)$, $(1, 6)$ and an asymptote with slope $\dfrac{2}{3}$.

    Answer: $\dfrac{(y - 4)^2}{4} - \dfrac{(x - 1)^2}{9} = 1$

    Difficulty: 2   Section: 3

51. Sketch the hyperbola $x^2 - 4y^2 + 6x - 16y + 9 = 0$ showing vertices, foci, and asymptotes on the graph.

    Answer: Foci are at $\left(-3, -3 \pm \sqrt{20}\right)$.

E172   Chapter 10   Exam Questions   Instructor's Resource Manual

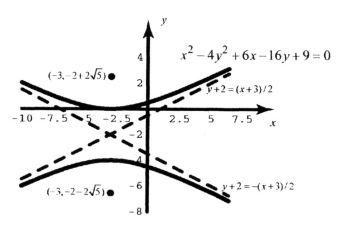

Difficulty: 2   Section: 3

52. Sketch the hyperbola $x^2 - 16y^2 - 4x + 32y - 76 = 0$ showing vertices, foci, and asymptotes on the graph.

   Answer: Foci are at $\left(2 \pm \sqrt{68}, 1\right)$.

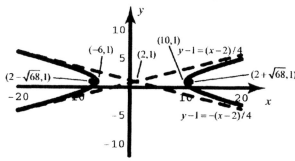

Difficulty: 2   Section: 3

53. Sketch the hyperbola $4x^2 - y^2 - 8x = 0$ showing vertices, foci, and asymptotes on the graph.

   Answer: Foci are at $\left(1 \pm \sqrt{5}, 0\right)$.

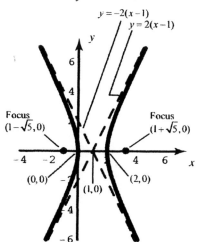

Difficulty: 2   Section: 3

54. Let θ be the angle of rotation to eliminate the xy term from the equation $4x^2 - 7xy = 4y^2 - 5x - 15$. Find the value of $\cot 2\theta$.

    Answer: $-\dfrac{8}{7}$

    Difficulty: 1   Section: 3

55. Find the angle θ so that the rotation through that angle will eliminate the xy term from the equation $x^2 - 2\sqrt{3}xy + 3y^2 - 16\sqrt{3}x - 16y = 0$.

    Answer: $\dfrac{\pi}{6}$

    Difficulty: 1   Section: 3

56. Let θ be the angle so that the rotation through θ will eliminate the xy term from the equation $4x^2 - 6xy - 4y^2 - 15 = 0$. Find $\sin\theta$ and $\cos\theta$.

    Answer: $\sin\theta = \dfrac{3}{\sqrt{10}}$, $\cos\theta = \dfrac{1}{\sqrt{10}}$

    Difficulty: 1   Section: 3

57. Write the equation of the curve $x^2 + y^2 - xy = 4$ in terms of u and v after rotating through the appropriate axes through the angle $\theta = \dfrac{\pi}{4}$.

    Answer: $u^2 + 3v^2 = 8$

    Difficulty: 1   Section: 3

58. Write the equation of the curve $8x^2 - 4xy + 5y^2 = 36$ in terms of u and v after rotating through the appropriate angle to eliminate the xy term.

    Answer: $\dfrac{u^2}{9} + \dfrac{v^2}{4} = 1$

    Difficulty: 1   Section: 3

59. Write the equation of the curve $16x^2 - 24xy + 9y^2 - 60x - 80y + 100$ in terms of u and v after rotating through the appropriate angle to eliminate the xy term.

    Answer: $v^2 = 4(u - 1)$

    Difficulty: 2   Section: 3

60. Write the equation of the curve $4x^2 - 6xy - 4y^2 - 15 = 0$ in terms of u and v after rotating through the appropriate angle to eliminate the xy term.

    Answer: $\dfrac{v^2}{3} - \dfrac{u^2}{3} = 1$

    Difficulty: 1   Section: 3

61. Use rotation of axes to eliminate the xy term and graph the conic $5x^2 + 6xy + 5y^2 - 8 = 0$. Identify the important parts.

Answer: $\dfrac{u^2}{1} + \dfrac{v^2}{4} = 1$

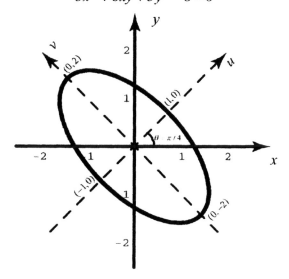

Difficulty: 2   Section: 3

62. Use rotation of axes to eliminate the xy term and graph the conic $5x^2 - 8xy + 5y^2 - 9 = 0$. Identify the important parts.

Answer: $\dfrac{u^2}{9} + \dfrac{v^2}{1} = 1$

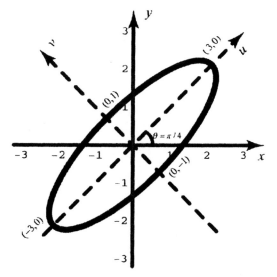

Difficulty: 2   Section: 3

63. Use rotation of axes to eliminate the xy term and graph the conic $2xy - 1 = 0$. Identify the

important parts.

Answer: $u^2 - v^2 = 1$

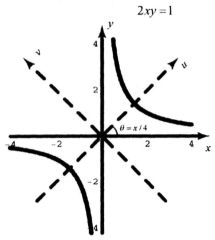

Difficulty: 2   Section: 3

64. Use rotation of axes to eliminate the xy term and graph the conic $x^2 - 2\sqrt{3}xy + 3y^2 - 16\sqrt{3}x - 16y = 0$. Identify the important parts.

Answer: $v^2 = 8u$

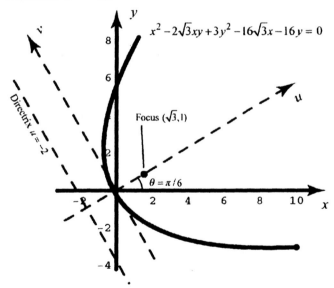

Difficulty: 2   Section: 3

65. Use rotation of axes to eliminate the xy term and graph the conic $5x^2 - 6xy + 5y^2 - 8x - 8y = 0$. Identify the important parts.

Answer: $\dfrac{\left(u - 2\sqrt{2}\right)^2}{8} + \dfrac{v^2}{2} = 1$

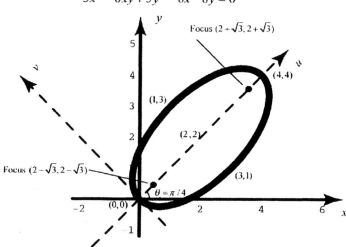

Difficulty: 3   Section: 3

66. Eliminate the parameter and identify the curve $x = \sqrt{1+t}$, $y = \sqrt{t}$, $t \geq 0$.

    Answer: $x^2 - y^2 = 1$, $x \geq 1$, $y \geq 1$, part of the hyperbola.

    Difficulty: 1   Section: 4

67. Eliminate the parameter and identify the curve $x = 3t^2$, $y = 2t - 1$, $-1 \leq t \leq 1$.

    Answer: $4x = 3\left(y^2 + 2y + 1\right)$, part of the parabola with $0 \leq x \leq 3$.

    Difficulty: 1   Section: 4

68. Eliminate the parameter and identify the curve $y = \sqrt{1 - \cos^2 t}$, $x = 1 - \sin^2 t$.

    Answer: $y^2 = 1 - x$, part of the parabola with $0 \leq x$, $0 \leq y$.

    Difficulty: 1   Section: 4

69. Eliminate the parameter and identify the curve $x = \sec t$, $y = \tan t$, $-\dfrac{\pi}{2} < t < \dfrac{\pi}{2}$.

    Answer: $x^2 - y^2 = 1$, part of the hyperbola with $x \geq 0$.

    Difficulty: 1   Section: 4

70. Eliminate the parameter and identify the curve $x = 2\sin\theta$, $y = 3\cos\theta$, $0 \leq \theta \leq \pi$.

    Answer: $\dfrac{x^2}{4} + \dfrac{y^2}{9} = 1$, the right half of an ellipse.

    Difficulty: 1   Section: 4

71. Find $\dfrac{dy}{dx}$ if $x = \sec t$ and $y = \tan t$, $-\dfrac{\pi}{2} < t < \dfrac{\pi}{2}$.

    Answer: $\dfrac{\sec t}{\tan t}$ if $t \neq 0$.

Difficulty: 1   Section: 4

72. Find $\dfrac{dy}{dx}$ if $x = 2\sin t$ and $y = 3\cos t$.

    Answer: $-\dfrac{3}{2}\tan t$

    Difficulty: 1   Section: 4

73. Find $\dfrac{d^2y}{dx^2}$ if $x = 2\sin t$ and $y = 3\cos t$.

    Answer: $-\dfrac{3}{4}\sec^3 t$

    Difficulty: 1   Section: 4

74. Find $\dfrac{dy}{dx}$ if $x = 4t^2$ and $y = t^3 - 12t$.

    Answer: $\dfrac{3t^2 - 12}{8t}$

    Difficulty: 1   Section: 4

75. Find $\dfrac{d^2y}{dx^2}$ if $x = 4t^2$ and $y = t^3 - 12t$, .

    Answer: $\dfrac{3t^2 + 12}{64t^3}$

    Difficulty: 1   Section: 4

76. The polar coordinates of P are $\left(4, \dfrac{\pi}{3}\right)$. Give the rectangular coordinates of P.

    Answer: $\left(2, 2\sqrt{3}\right)$

    Difficulty: 1   Section: 5

77. The polar coordinates of P are $\left(2, \dfrac{7\pi}{6}\right)$. Give the rectangular coordinates of P.

    Answer: $\left(-\sqrt{3}, -1\right)$

    Difficulty: 1   Section: 5

78. The polar coordinates of P are $(4, \pi)$. Give the rectangular coordinates of P.

    Answer: $(-4, 0)$

    Difficulty: 1   Section: 5

79. The polar coordinates of P are $\left(3, -\dfrac{\pi}{4}\right)$. Give the rectangular coordinates of P.

Answer: $\left(\dfrac{3\sqrt{2}}{2}, -\dfrac{3\sqrt{2}}{2}\right)$

Difficulty: 1   Section: 5

80. The rectangular coordinates of P are $(2,2)$. Give the polar coordinates of P.

    Answer: $\left(2\sqrt{2}, \dfrac{\pi}{4} + 2n\pi\right)$ or $\left(-2\sqrt{2}, \dfrac{\pi}{4} + (2n+1)\pi\right)$

    Difficulty: 1   Section: 5

81. The rectangular coordinates of P are $(0,3)$. Give the polar coordinates of P.

    Answer: $\left(3, \dfrac{\pi}{2} + 2n\pi\right)$ or $\left(-3, \dfrac{3\pi}{2} + 2n\pi\right)$

    Difficulty: 1   Section: 5

82. The rectangular coordinates of P are $(0,3)$. Give the polar coordinates of P.

    Answer: $\left(8, \dfrac{7\pi}{6} + 2n\pi\right)$ or $\left(-8, \dfrac{7\pi}{6} + (2n+1)\pi\right)$

    Difficulty: 1   Section: 5

83. The polar coordinates of P are $\left(-3, \dfrac{\pi}{2}\right)$. Give the rectangular coordinates of P.

    Answer: $(0, -3)$

    Difficulty; 1   Section: 5

84. The polar coordinates of P are $\left(-2, \dfrac{\pi}{3}\right)$. Give the rectangular coordinates of P.

    Answer: $(-1, \sqrt{3})$

    Difficulty; 1   Section: 5

85. The rectangular coordinates of P are $(-2, -3)$. Give the polar coordinates of P.

    Answer: $\left(\sqrt{13}, \tan^{-1}(1.5) + (2n+1)\pi\right)$ or $\left(-\sqrt{13}, \tan^{-1}(1.5) + (2n+1)\pi\right)$

    Difficulty: 1   Section: 5

86. The rectangular coordinates of P are $(1, -4)$. Give the polar coordinates of P.

    Answer: $\left(\sqrt{17}, \tan^{-1}(-4) + 2n\pi\right)$ or $\left(-\sqrt{17}, \tan^{-1}(-4) + 2n\pi\right)$

    Difficulty: 1   Section: 5

87. The rectangular coordinates of P are $(-3, -3)$. Give the polar coordinates of P.

    Answer: $\left(3\sqrt{2}, \dfrac{5\pi}{4} + 2n\pi\right)$ or $\left(-3\sqrt{2}, \dfrac{\pi}{4} + 2n\pi\right)$

Difficulty: 1   Section: 5

88. Find a polar equation for the graph of the Cartesian equation $x^2 + y^2 - 2x = 9$.

    Answer: $r^2 = 2r(\cos\theta) + 9$

    Difficulty: 1   Section: 5

89. Find a polar equation for the graph of the Cartesian equation $y^2 = 4x$

    Answer: $r(\sin^2) = 4\cos\theta$

    Difficulty: 1   Section: 5

90. Find a Cartesian equation for the graph of the polar equation $r = 3\cos\theta$.

    Answer: $x^2 + y^2 = 3y$.

    Difficulty: 1   Section: 5

91. Find a Cartesian equation for the graph of the polar equation $r^2 = 2r(\cos\theta) + r(\sin\theta)$.

    Answer: $x^2 + y^2 = 2x + y$

    Difficulty: 1   Section: 5

92. Name the curve that has the polar equation $r = 2\cos\theta$.

    Answer: circle

    Difficulty: 1   Section: 5

93. Name the curve that has the polar equation $r = \dfrac{5}{3 + 2\cos\theta}$ and give its eccentricity.

    Answer: ellipse, $e = \dfrac{2}{3}$

    Difficulty: 2   Section: 5

94. Name the curve that has the polar equation $f = \dfrac{5}{2 + 3\cos\theta}$ and give its eccentricity.

    Answer: hyperbola, $e = \dfrac{3}{2}$

    Difficulty: 2   Section: 5

95. Name the curve that has the polar equation $r = \dfrac{5}{3 - 3\cos\theta}$ and give its eccentricity.

    Answer: parabola, $e = 1$

    Difficulty: 2   Section: 5

96. Sketch the graph of $r = 3\sec\theta$.

Answer: The graph is the line $x = 3$.

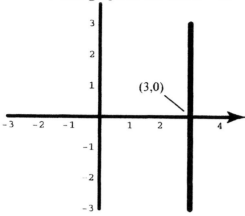

Difficulty: 1   Section: 6

97. Sketch the graph of $r = -3\cos\theta$.

    Answer: The graph is a circle of radius $\dfrac{3}{2}$ with center at $\left(\dfrac{3}{2}, \pi\right)$.

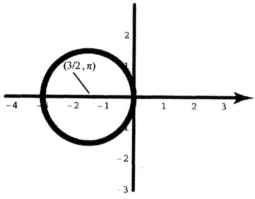

Difficulty: 1   Section: 6

98. Sketch the graph of $r = 6\sin\theta$.

    Answer: The graph is a circle of radius 3, with center at $\left(3, \dfrac{\pi}{2}\right)$.

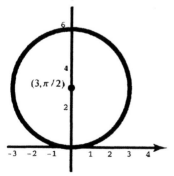

Difficulty: 1   Section: 6

99. Sketch the graph of $r = 3\sin 3\theta$.

Answer: The graph is a 3-leafed rose.

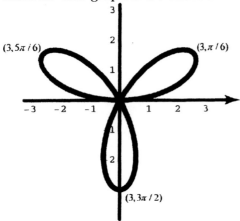

Difficulty: 2   Section: 6

100. Sketch the graph of $r^2 = 4\sin 2\theta$.

Answer: The graph is a lemniscate.

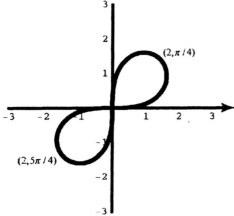

Difficulty: 2   Section: 6

101. Sketch the graph of $r = 1 + 2\sin\theta$.

Answer: The graph is a limacon.

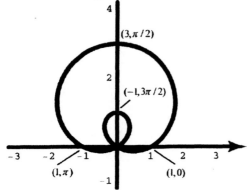

Difficulty: 2   Section: 6

102. Sketch the graph of $r = 2 + 2\sin\theta$.

Answer: The graph is a cardioid.

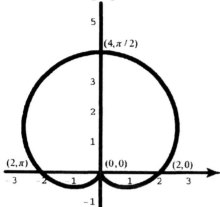

Difficulty: 2   Section: 6

103. Find the points of intersection of the curves $r = 2$ and $r = 4\sin 2\theta$ that are in quadrant I.

Answer: $\left(2, \dfrac{\pi}{12}\right)$ and $\left(2, \dfrac{5\pi}{12}\right)$

Difficulty: 2   Section: 6

104. Find the points of intersection of the curves $r = 2\sin\theta$ and $r = 2\cos\theta$.

Answer: $\left(\sqrt{2}, \dfrac{\pi}{4}\right)$ and $(0,0)$

Difficulty: 2   Section: 6

105. Find the area bounded by the curve $r^2 = 2\cos 2\theta + 1$.

Answer: $\sqrt{3} + \dfrac{2\pi}{3}$

Difficulty: 1   Section: 7

106. Find the area bounded by the curve $r^2 = \dfrac{1}{2}\sin 2\theta$.

Answer: $\dfrac{1}{2}$

Difficulty: 1   Section: 7

107. Find the area bounded by the curve $r^2 = 2\sin 3\theta$.

Answer: 4

Difficulty: 1   Section: 7

108. Find the area bounded by the curve $r^2 = 2\cos 3\theta$.

Answer: 4

Difficulty: 1   Section: 7

109. Find the area inside the circle $r = 3\cos\theta$ and outside the cardioid $r = 1 + \cos\theta$.

   Answer: π

   Difficulty: 2   Section: 7

110. Find the area inside the circle $r = 3\sin\theta$ and outside the cardioid $r = 1 + \sin\theta$.

   Answer: π

   Difficulty: 2   Section: 7

111. Find the area inside the cardioid $r = 1 + \cos\theta$.

   Answer: $\dfrac{3\pi}{2}$

   Difficulty: 1   Section: 7

112. Find the area inside the cardioid $r = 1 + \sin\theta$.

   Answer: $\dfrac{3\pi}{2}$

   Difficulty: 2   Section: 7

113. Find the area inside the cardioid $r = 2\cos 3\theta$.

   Answer: $\dfrac{\pi}{3}$

   Difficulty: 2   Section: 7

114. Find the area inside one petal of the rose $r = 2\sin 3\theta$.

   Answer: $\dfrac{\pi}{3}$

   Difficulty: 2   Section: 7

115. Find the length of the cardioid $r = 1 + \cos\theta$.

   Answer: 8

   Difficulty: 2   Section: 7

116. Find the length of the cardioid $r = 1 + \sin 2\theta$.

   Answer: 8

   Difficulty: 2   Section: 7

117. Find the length of the curve $r = \sin^3\left(\dfrac{\theta}{3}\right)$.

Answer: $\dfrac{3\pi}{2}$

Difficulty: 2   Section: 7

118. Find the equation of the tangent line to $r = 4\cos 2\theta$ at $\theta = \dfrac{\pi}{2}$.

   Answer: $y = -4$

   Difficulty: 1   Section: 7

119. Find the slope of the tangent line to the circle $r = 2\sin\theta$ at the point where $\theta = \dfrac{\pi}{3}$.

   Answer: $-\sqrt{3}$

   Difficulty: 1   Section: 7

120. Find the slope of the cardioid $r = 1 - \cos\theta$ at $\theta = \dfrac{\pi}{2}$.

   Answer: -1

   Difficulty: 1   Section: 7

121. Find all points where the cardioid $r = 1 + \sin\theta$ has a horizontal tangent line.

   Answer: $\left(2, \dfrac{\pi}{2}\right)$, $\left(\dfrac{1}{2}, \dfrac{7\pi}{6}\right)$, $\left(\dfrac{1}{2}, \dfrac{11\pi}{6}\right)$

   Difficulty: 2   Section: 7

122. Find all points where the cardioid $r = 1 + \sin\theta$ has a vertical tangent line.

   Answer: $\left(0, \dfrac{3\pi}{2}\right)$, $\left(\dfrac{3}{2}, \dfrac{\pi}{6}\right)$, $\left(\dfrac{3}{2}, \dfrac{5\pi}{6}\right)$

   Difficulty: 2   Section: 7

11 Geometry in Space and Vectors

1. Find the distance from $(0, 1, 2)$ to $(1, -2, 3)$.
   Answer: $\sqrt{35}$
   Difficulty: 1   Section: 1

2. Find the distance from $(1, -4, 5)$ to $(2, 0, 3)$.
   Answer: $\sqrt{21}$
   Difficulty: 1   Section: 1

3. Find the distance from $(-4, 3, -2)$ to $(-2, 1, 0)$.
   Answer: $\sqrt{12}$
   Difficulty: 1   Section: 1

4. Find the midpoint of the line segment joining $(0, 1, -2)$ and $(1, -2, 3)$.
   Answer: $\left(\dfrac{1}{2}, -\dfrac{1}{2}, \dfrac{1}{2}\right)$
   Difficulty: 1   Section: 1

5. Find the midpoint of the line segment joining $(1, -4, 5)$ and $(2, 0, 3)$.
   Answer: $\left(\dfrac{3}{2}, -2, 4\right)$
   Difficulty: 1   Section: 1

6. Find the midpoint of the line segment joining $(-4, 3, -2)$ and $(-2, 1, 0)$.
   Answer: $\left(-3, \dfrac{3}{2}, -1\right)$
   Difficulty: 1   Section: 1

7. Find the center and radius of the sphere with equation $x^2 - 2x + y^2 + 6y + z^2 = 6$.
   Answer: center $(1, -3, 0)$, radius 4
   Difficulty: 1   Section: 1

8. Find the equation of the sphere with center $(1, -3, 0)$ and radius 4.
   Answer: $(x - 1)^2 + (y + 3)^2 + z^2 = 16$
   Difficulty: 1   Section: 1

9. Find the center and radius of the sphere with equation $x^2 + 6x + y^2 - 4x + z^2 - 8z = -4$.

Answer: center $(-3, 2, 4)$, radius 5

Difficulty: 1   Section: 1

10. Find the equation of the sphere with center $(-3, 2, 4)$ and radius 5.

    Answer: $(x+3)^2 + (y-2)^2 + (z-4)^2 = 25$

    Difficulty: 1   Section: 1

11. An airplane flies east with an airspeed in still air of 200 mph. A wind is blowing E 60° N at 120 mph. What is the velocity of the airplane relative to the ground and what is its direction of travel?

    Answer: 279.99 mph, E 21.78° N

    Difficulty: 2   Section: 2

12. An airplane flies east with an airspeed in still air of 200 mph. A wind is blowing E 30° N at 120 mph. What is the velocity of the airplane relative to the ground and what is its direction of travel?

    Answer: 309.77 mph, E 11.17° N

    Difficulty: 2   Section: 2

13. A motorboat travels at a speed of 30 mph in still water. The boat starts upriver at an angle of 40° with the riverbank against a 5 mph current. At what angle with the riverbank is the boat moving and at what speed relative to the riverbank?

    Answer: 47.77°, 26.4 mph

    Difficulty: 2   Section: 2

14. A river flows south with a current of 5 mph. A motorboat travels at a speed of 25 mph in still water. In what direction must the boat be headed in order to move directly east across the river?

    Answer: 78.5°

    Difficulty: 2   Section: 2

15. An airplane is flying at a heading of 40° (40° east of north) with an airspeed of 600 mph. The wind is blowing from the south at 25 mph. What is the direction in which the plane flies and what is its ground speed?

    Answer: A heading of 39° (39° east of north) with an airspeed of 620 mph

    Difficulty: 2   Section: 2

16. If $\mathbf{u} = 2\mathbf{i} + 3\mathbf{j}$ and $\mathbf{v} = -\mathbf{i} - 2\mathbf{j}$, find $\mathbf{u} + \mathbf{v}$.

    Answer: $\mathbf{i} + \mathbf{j}$

Difficulty: 1   Section: 2

17. If $\mathbf{u} = 3\mathbf{i} - \mathbf{j}$ and $\mathbf{v} = 2\mathbf{i} - 3\mathbf{j}$, find $2\mathbf{u} - \mathbf{v}$.

    Answer: $4\mathbf{i} + \mathbf{j}$

    Difficulty: 1   Section: 2

18. If $\mathbf{u} = 4\mathbf{i} - 3\mathbf{j}$, compute the length of $\mathbf{u}$.

    Answer: 5

    Difficulty: 1   Section: 2

19. f $\mathbf{u} = \mathbf{i} - 3\mathbf{j} + 2\mathbf{k}$, compute the length of $\mathbf{u}$.

    Answer: $\sqrt{10}$

    Difficulty: 1   Section: 2

20. Find the vector extending from $(0, 2)$ to $(2, -4)$.

    Answer: $2\mathbf{i} - 6\mathbf{j}$

    Difficulty: 1   Section: 2

21. Find the vector extending from $(1, -2)$ to $(1, 2)$.

    Answer: $4\mathbf{j}$

    Difficulty: 1   Section: 2

22. Find the vector extending from $(-2, 3)$ to $(-1, 3)$.

    Answer: $\mathbf{i}$

    Difficulty: 1   Section: 2

23. Find a vector of length 4 in the direction $\mathbf{i} - \mathbf{j}$.

    Answer: $\dfrac{4}{\sqrt{2}}(\mathbf{i} - \mathbf{j})$

    Difficulty: 1   Section: 2

24. Find a vector of length 2 in the direction of $2\mathbf{i} - \mathbf{j}$.

    Answer: $\dfrac{2}{\sqrt{5}}(2\mathbf{i} - \mathbf{j})$

    Difficulty: 1   Section: 2

25. Find the vector of length 5 in the direction of $4\mathbf{i} - 3\mathbf{j}$.

    Answer: $4\mathbf{i} - 3\mathbf{j}$

Difficulty: 1   Section: 2

26. Find a vector of length 5 in the direction of $3\mathbf{i} - 4\mathbf{j}$.

    Answer: $3\mathbf{i} - 4\mathbf{j}$

    Difficulty: 1   Section: 2

27. If $\mathbf{u} = 2\mathbf{i} + 3\mathbf{j} - \mathbf{k}$ and $\mathbf{v} = -\mathbf{i} - 2\mathbf{j}$, find $\mathbf{u} + \mathbf{v}$.

    Answer: $\mathbf{i} + \mathbf{j} - \mathbf{k}$

    Difficulty: 1   Section: 2

28. If $\mathbf{u} = 3\mathbf{i} - \mathbf{j} + 9\mathbf{k}$ and $\mathbf{v} = 2\mathbf{i} - 3\mathbf{j} + \mathbf{k}$, find $2\mathbf{u} - \mathbf{v}$.

    Answer: $4\mathbf{i} + \mathbf{j} + 17\mathbf{k}$

    Difficulty: 1   Section: 2

29. If $\mathbf{u} = 4\mathbf{i} - 3\mathbf{j} - \mathbf{k}$, compute the length of $\mathbf{u}$.

    Answer: $\sqrt{26}$

    Difficulty: 1   Section: 2

30. If $\mathbf{u} = \mathbf{i} - 3\mathbf{j} - 2\mathbf{k}$, compute the length of $\mathbf{u}$.

    Answer: $\sqrt{14}$

    Difficulty: 1   Section: 2

31. Write the vector extending from $(0, 2, 3)$ to $(2, -4, -1)$.

    Answer: $2\mathbf{i} - 6\mathbf{j} - 4\mathbf{k}$

    Difficulty: 1   Section: 2

32. Write the vector extending from $(1, -2, 4)$ to $(1, 2, -1)$.

    Answer: $4\mathbf{j} - 5\mathbf{k}$

    Difficulty: 1   Section: 2

33. Write the vector extending from $(-2, 3, -1)$ to $(-1, 3, 4)$.

    Answer: $\mathbf{i} + 5\mathbf{k}$

    Difficulty: 1   Section: 2

34. Find a vector of length 4 in the direction of $\mathbf{i} - \mathbf{j} + 2\mathbf{k}$.

    Answer: $\dfrac{4}{\sqrt{6}}(\mathbf{i} - \mathbf{j} + 2\mathbf{k})$

    Difficulty: 1   Section: 2

35. Find a vector of length 2 in the direction of $2\mathbf{i} - \mathbf{j} - 3\mathbf{k}$.

    Answer: $\dfrac{2}{\sqrt{14}}(2\mathbf{i} - \mathbf{j} - 3\mathbf{k})$

    Difficulty: 1   Section: 2

36. Find a vector of length 5 in the direction of $4\mathbf{i} - 3\mathbf{j}$.

    Answer: $4\mathbf{i} - 3\mathbf{j}$

    Difficulty: 1   Section: 2

37. Find the vector of length 5 in the direction of $3\mathbf{j} - 4\mathbf{k}$.

    Answer: $3\mathbf{j} - 4\mathbf{k}$

    Difficulty: 1   Section: 2

38. If the length of $\mathbf{u}$ is 4 and the length of $\mathbf{v}$ is 3, what can be said about the length of $\mathbf{u} + \mathbf{v}$?

    Answer: The length of $\mathbf{u} + \mathbf{v}$ is less than or equal to 7.

    Difficulty: 1   Section: 2

39. If the length of $\mathbf{u}$ is 3 and the length of $\mathbf{v}$ is 5, what can be said about the length of $\mathbf{u} + \mathbf{v}$?

    Answer: The length of $\mathbf{u} + \mathbf{v}$ is less than or equal to 8.

    Difficulty: 1   Section: 2

40. If the length of $\mathbf{u}$ is 4, what is the length of $-2\mathbf{u}$?

    Answer: 8

    Difficulty: 1   Section: 2

41. If the length of $\mathbf{u}$ is 2, what is the length of $-\mathbf{u}$?

    Answer: $\sqrt{2}$

    Difficulty: 1   Section: 2

42. Find a unit vector in the direction of $2\mathbf{i} - \mathbf{j} + \mathbf{k}$.

    Answer: $\dfrac{1}{\sqrt{6}}(2\mathbf{i} - \mathbf{j} + \mathbf{k})$

    Difficulty: 1   Section: 2

43. Find a unit vector in the direction opposite to that of $3\mathbf{i} - 3\mathbf{j} + \mathbf{k}$.

    Answer: $\dfrac{1}{\sqrt{19}}(-3\mathbf{i} + 3\mathbf{j} - \mathbf{k})$

    Difficulty: 1   Section: 2

44. Find a unit vector in the direction opposite to that of $2\mathbf{i} - 3\mathbf{j} + 3\mathbf{k}$.

    Answer: $\dfrac{1}{\sqrt{22}}(-2\mathbf{i} + 3\mathbf{j} + 3\mathbf{k})$

    Difficulty: 1   Section: 2

45. Find a vector of length $\sqrt{2}$ in the direction of the vector extending from $(2, -1, 0)$ to $(-1, 2, 2)$.

    Answer: $\dfrac{\sqrt{2}}{\sqrt{22}}(-3\mathbf{i} + 3\mathbf{j} + 2\mathbf{k})$

    Difficulty: 1   Section: 2

46. The vector $3\mathbf{i} - 2\mathbf{j} + 4\mathbf{k}$ extends from $(1, -4, -5)$ to (a, b, c). What are the coordinates of a, b, and c?

    Answer: $a = 4$, $b = -6$, $c = -1$

    Difficulty: 2   Section: 2

47. Find the angle between the vectors $\sqrt{3}\mathbf{i} - \mathbf{j}$ and $-\mathbf{i} + \sqrt{3}\mathbf{j}$.

    Answer: $\dfrac{5\pi}{6}$

    Difficulty: 1   Section: 3

48. Find the cosine of the angle between $\mathbf{i} - 4\mathbf{j}$ and $4\mathbf{i} + 2\mathbf{j}$.

    Answer: $\dfrac{-2}{\sqrt{85}}$

    Difficulty: 1   Section: 3

49. Find the dot product of $2\mathbf{i} - \mathbf{k}$ and $3\mathbf{i} + 14\mathbf{j} - \mathbf{k}$.

    Answer: 7

    Difficulty: 1   Section: 3

50. Find the dot product of $4\mathbf{i} - \mathbf{j} + 5\mathbf{k}$ and $2\mathbf{i} + \mathbf{j} - \mathbf{k}$.

    Answer: 2

    Difficulty: 1   Section: 3

51. Find the dot product of $2\mathbf{i} - 3\mathbf{j} + \mathbf{k}$ and $2\mathbf{i} + \mathbf{j} - \mathbf{k}$.

    Answer: 0

    Difficulty: 1   Section: 3

52. Find the dot product of $3\mathbf{i} + \mathbf{j} + 2\mathbf{k}$ and $\mathbf{i} + \mathbf{j} - 2\mathbf{k}$.

    Answer: 0

Difficulty: 1   Section: 3

53. Find the dot product of $4\mathbf{i} - \mathbf{j} - \mathbf{k}$ and $\mathbf{i} + 5\mathbf{j} + \mathbf{k}$.

    Answer: -2

    Difficulty: 1   Section: 3

54. Find the dot product of $\mathbf{i} - \mathbf{j} - \mathbf{k}$ and $3\mathbf{i} + 5\mathbf{j} + \mathbf{k}$.

    Answer: -3

    Difficulty: 1   Section: 3

55. Find the angle between the vectors $\sqrt{3}\mathbf{i} - \mathbf{j}$ and $-\mathbf{i} + \sqrt{3}\mathbf{j}$.

    Answer: $\dfrac{5\pi}{6}$

    Difficulty: 1   Section: 3

56. Find the cosine of the angle between $\mathbf{i} - 4\mathbf{j}$ and $4\mathbf{i} - 2\mathbf{j}$.

    Answer: $\dfrac{-2}{\sqrt{85}}$

    Difficulty: 1   Section: 3

57. Find the cosine of the angle between $4\mathbf{i} - 3\mathbf{j} + \mathbf{k}$ and $\mathbf{i} + 2\mathbf{j} - \mathbf{k}$.

    Answer: $\dfrac{4}{3\sqrt{26}}$

    Difficulty: 1   Section: 3

58. Find the cosine of the angle between $-2\mathbf{i} + \mathbf{j} + 3\mathbf{k}$ and $\mathbf{i} - \mathbf{j} - 2\mathbf{k}$.

    Answer: $-\dfrac{9}{\sqrt{84}}$

    Difficulty: 1   Section: 3

59. Find the equation of the plane containing $(2, -3, 1)$ and perpendicular to the vector $\mathbf{i} - \mathbf{j} + 9\mathbf{k}$.

    Answer: $(x - 2) - (y + 3) + 9(z - 1) = 0$

    Difficulty: 1   Section: 3

60. Find the equation of the plane containing $(1, -1, 9)$ and perpendicular to the vector $2\mathbf{i} - 3\mathbf{j} + \mathbf{k}$.

    Answer: $2(x - 1) - 3(y + 1) + (z - 9) = 0$

    Difficulty: 1   Section: 3

61. Find the equation of the plane containing $(1, -1, 3)$ and perpendicular to the vector $3\mathbf{i} - \mathbf{j}$.

Answer: $3(x-1) - (y+1) = 0$

Difficulty: 1   Section: 3

62. Find the equation of the plane containing $(1, -1, 3)$ and perpendicular to the vector $3\mathbf{j} - \mathbf{k}$.

Answer: $3(y+1) - (z-3) = 0$

Difficulty: 1   Section: 3

63. Find the equation of the plane containing $(3, 1, 0)$ and perpendicular to the vector $4\mathbf{i} - 5\mathbf{j} + 8\mathbf{k}$.

Answer: $4(x-3) - 5(y-1) + 8z = 0$

Difficulty: 1   Section: 3

64. If $\mathbf{u} = 2\mathbf{i} - \mathbf{j} - \mathbf{k}$ and $\mathbf{v} = \mathbf{i} + 3\mathbf{j} + 3\mathbf{k}$, find $pr_{\mathbf{v}}\mathbf{u}$.

Answer: $\dfrac{-4}{19}(\mathbf{i} + 3\mathbf{j} + 3\mathbf{k})$

Difficulty: 2   Section: 3

65. If $\mathbf{u} = \mathbf{i} + \mathbf{j} - 8\mathbf{k}$ and $\mathbf{v} = 2\mathbf{i} + 2\mathbf{j} - \mathbf{k}$, find $pr_{\mathbf{v}}\mathbf{u}$.

Answer: $\dfrac{12}{9}(2\mathbf{i} + 2\mathbf{j} - \mathbf{k})$

Difficulty: 2   Section: 3

66. If $\mathbf{u} = 2\mathbf{i} + 2\mathbf{j} - \mathbf{k}$ and $\mathbf{v} = \mathbf{i} + \mathbf{j} - 8\mathbf{k}$, find $pr_{\mathbf{v}}\mathbf{u}$.

Answer: $\dfrac{12}{66}(\mathbf{i} + \mathbf{j} - 8\mathbf{k})$

Difficulty: 2   Section: 3

67. If $\mathbf{u} = \mathbf{i} + 3\mathbf{j} + 3\mathbf{k}$ and $\mathbf{v} = 2\mathbf{i} - \mathbf{j} - \mathbf{k}$, find $pr_{\mathbf{v}}\mathbf{u}$.

Answer: $\dfrac{-4}{6}(2i - j - k)$

Difficulty: 2   Section: 3

68. If $\mathbf{u} = \mathbf{i} - 2\mathbf{j} - \mathbf{k}$ and $\mathbf{v} = 2\mathbf{i} + 2\mathbf{j} - 2\mathbf{k}$, find $pr_{\mathbf{v}}\mathbf{u}$.

Answer: the zero vector

Difficulty: 2   Section: 3

69. If $\mathbf{u} = 3\mathbf{i} + 2\mathbf{j} - 4\mathbf{k}$ and $\mathbf{v} = 2\mathbf{i} - 5\mathbf{j} - \mathbf{k}$, find $pr_{\mathbf{v}}\mathbf{u}$.

Answer: the zero vector

Difficulty: 2   Section: 3

70. If $\mathbf{u} = 2\mathbf{i} - \mathbf{j} + 6\mathbf{k}$ and $\mathbf{v} = -3\mathbf{i} + 5\mathbf{j} + 5\mathbf{k}$, find $\mathbf{u} \times \mathbf{v}$.

Answer: $-31i - 20j + 7k$

Difficulty: 1   Section: 4

71. If $\mathbf{u} = -2\mathbf{i} + \mathbf{j} + 3\mathbf{k}$ and $\mathbf{V} = \mathbf{i} - \mathbf{j} - 2\mathbf{k}$, find $\mathbf{u} \times \mathbf{v}$.

    Answer: $i - j + k$

    Difficulty: 1   Section: 4

72. If $\mathbf{u} = 2\mathbf{i} - \mathbf{j} + 3\mathbf{k}$ and $\mathbf{v} = \mathbf{i} - 2\mathbf{j} - \mathbf{k}$, find $\mathbf{u} \times \mathbf{v}$.

    Answer: $7\mathbf{i} + 5\mathbf{j} - 3\mathbf{k}$

    Difficulty: 1   Section: 4

73. If $\mathbf{u} = 3\mathbf{i} - \mathbf{j} - 4\mathbf{k}$ and $\mathbf{v} = 2\mathbf{i} + 5\mathbf{j} - 2\mathbf{k}$, find $\mathbf{u} \times \mathbf{v}$.

    Answer: $22\mathbf{i} - 2\mathbf{j} + 17\mathbf{k}$

    Difficulty: 1   Section: 4

74. If $\mathbf{u} = 2\mathbf{i} - \mathbf{j} + 3\mathbf{k}$ and $\mathbf{v} = \mathbf{i} + \mathbf{j} - 4\mathbf{k}$, find $\mathbf{u} \times \mathbf{v}$.

    Answer: $\mathbf{i} + 11\mathbf{j} + 3\mathbf{k}$

    Difficulty: 1   Section: 4

75. Find the equation of the plane through $(1, 2, -3)$, $(3, 1, 0)$, and $(2, 3, -7)$.

    Answer: $x + 11y + 3z - 14 = 0$

    Difficulty: 2   Section: 4

76. Find the equation of the plane through $(1, 2, -1)$, $(3, 1, 2)$, and $(2, 3, -5)$.

    Answer: $x + 11y + 3z - 20$

    Difficulty: 2   Section: 4

77. Find the equation of the plane through $(1, 2, -1)$, $(3, -3, 7)$, and $(-2, 3, 2)$.

    Answer: $31x + 20y - 7z = -16$

    Difficulty: 2   Section: 4

78. Find the equation of the plane through $(6, 1, 0)$, $(5, -2, -1)$, and $(1, -3, 4)$.

    Answer: $16x - 9y + 11z = 87$

    Difficulty: 2   Section: 4

79. Find the equation of the plane through $(2, 3, 1)$, $(1, -1, 2)$, and $(-4, 3, 0)$.

    Answer: $4x - 7y - 24z = -37$

Difficulty: 2   Section: 4

80. Find the equation of the plane through $(2,3,1)$, $(1,-1,2)$, and $(-4,3,1)$.

    Answer: $y + 4z = 7$

    Difficulty: 2   Section: 4

81. Find a unit vector perpendicular to both $\mathbf{i} + 2\mathbf{j} - \mathbf{k}$ and $2\mathbf{i} - 2\mathbf{j} + 3\mathbf{k}$.

    Answer: $\dfrac{1}{\sqrt{77}}(4\mathbf{i} - 5\mathbf{j} - 6\mathbf{k})$

    Difficulty: 2   Section: 4

82. Find a unit vector perpendicular to both $-2\mathbf{i} + \mathbf{j} + 3\mathbf{k}$ and $\mathbf{i} - \mathbf{j} - 2\mathbf{k}$.

    Answer: $\dfrac{1}{\sqrt{3}}(\mathbf{i} - \mathbf{j} + \mathbf{k})$

    Difficulty: 2   Section: 4

83. Find a vector of length 2 perpendicular to both $2\mathbf{i} - \mathbf{j} + 3\mathbf{k}$ and $\mathbf{i} - 2\mathbf{j} - \mathbf{k}$.

    Answer: $\dfrac{2}{\sqrt{83}}(7\mathbf{i} + 5\mathbf{j} - 3\mathbf{k})$

    Difficulty: 2   Section: 4

84. If $\mathbf{r}(t) = t^2\mathbf{i} + \dfrac{t^4 - 4}{t - 2}\mathbf{j}$, find $\lim\limits_{t \to 2} \mathbf{r}(t)$.

    Answer: $4\mathbf{i} + 4\mathbf{j}$

    Difficulty: 1   Section: 5

85. If $\mathbf{r}(t) = 3t^2\mathbf{i} + \dfrac{t^2 - 9}{t - 3}\mathbf{j}$, find $\lim\limits_{t \to 3} \mathbf{r}(t)$.

    Answer: $27\mathbf{i} + 6\mathbf{j}$

    Difficulty: 1   Section: 5

86. If $\mathbf{r}(t) = \ln t\, \mathbf{i} + 2t\, \mathbf{j}$, find $\lim\limits_{t \to 0} \mathbf{r}(t)$ if it exists.

    Answer: limit does not exist

    Difficulty: 1   Section: 5

87. If $\mathbf{r}(t) = \dfrac{t+1}{t-1}\mathbf{i} + \ln(2t^3 + 1)\mathbf{j}$, find the velocity vector and acceleration vector when $t = 0$.

    Answer: $\mathbf{v} = -2\mathbf{i}$, $\mathbf{a} = -4\mathbf{i}$

    Difficulty: 1   Section: 5

88. If $\mathbf{r}(t) = \sec t\, \mathbf{i} + \tan t\, \mathbf{i}$, find the velocity vector and acceleration vector when $t = 0$.

    Answer:
    $$\mathbf{v} = (\sec t \tan t)\,\mathbf{i} + \sec^2 t\,\mathbf{j}$$
    $$\mathbf{a} = \left(\sec t \tan^2 t + \sec^3 t\right)\mathbf{i} + \left(2\sec^2 t \tan t\right)\mathbf{j}$$

    Difficulty: 1   Section: 5

89. Find the position and velocity vectors if $\mathbf{a} = 2t\mathbf{i} - \mathbf{j}$, $\mathbf{v}(0) = \mathbf{i}$, and $\mathbf{r}(0) = 2\mathbf{i} - \mathbf{j}$.

    Answer: $\mathbf{v}(t) = \left(t^2 + 1\right)\mathbf{i} - t\mathbf{j}$, $\mathbf{r}(t) = \left(\dfrac{t^3}{3} + t + 2\right)\mathbf{i} - \left(\dfrac{t^2}{2} + 1\right)\mathbf{j}$

    Difficulty: 1   Section: 5

90. Find the position and velocity vectors if $\mathbf{a} = \cos t\,\mathbf{i} + \sin^2 t\,\mathbf{j}$, $\mathbf{v}\left(\dfrac{\pi}{2}\right) = \dfrac{3}{2}\mathbf{j}$, $\mathbf{r}\left(\dfrac{\pi}{2}\right) = \dfrac{\pi}{2}\mathbf{j}$.

    Answer:
    $$\mathbf{v}(t) = (\sin t - 1)\mathbf{i} + \left(\dfrac{-\cos 2t}{2} + 1\right)\mathbf{j}$$
    $$\mathbf{r}(t) = \left(-\cos t - t + \dfrac{\pi}{2}\right)\mathbf{i} + \left(\dfrac{-\sin 2t}{4} + t\right)\mathbf{j}$$

    Difficulty: 1   Section: 5

91. Find the position and velocity vectors if $\mathbf{a} = 12t^2\mathbf{i} - (4\cos t)\mathbf{j}$, $\mathbf{v}(0) = -\mathbf{i}$, and $\mathbf{r}(0) = 2\mathbf{i} + \mathbf{j}$.

    Answer:
    $$\mathbf{v}(t) = \left(4t^3 - 1\right)\mathbf{i} - (4\sin t)\mathbf{j}$$
    $$\mathbf{r}(t) = \left(t^4 - t + 2\right)\mathbf{i} + (4\cos t - 3)\mathbf{j}$$

    Difficulty: 1   Section: 5

92. A shell is fired from ground level, with muzzle velocity of 600 ft/sec., and an angle of elevation of 30 degrees. Find the parametric equations of the trajectory.

    Answer: $x(t) = 300\sqrt{3}t$, $y = 300t - 16t^2$

    Difficulty: 1   Section: 5

93. A shell is fired from ground level, with muzzle velocity of 600 ft/sec., and an angle of elevation of 60 degrees. Find the parametric equations of the trajectory.

    Answer: $x(t) = 300t$, $y = 300\sqrt{3}t - 16t^2$

    Difficulty: 1   Section: 5

94. Find the velocity and acceleration vectors for $x\sin^2 t$, $y = \cos^2 t$.

    Answer:
    $$\mathbf{v} = (2\sin t \cos t)\,\mathbf{i} - (2\sin t \cos t)\,\mathbf{j}$$
    $$\mathbf{a} = 2\left(\cos^2 t - \sin^2 t\right)\mathbf{i} + 2\left(\sin^2 t - \cos^2 t\right)\mathbf{j}$$

    Difficulty: 1   Section: 5

95. Find the velocity and acceleration vectors for $\dfrac{t+1}{t-1}$ and $y = \ln\left(2t^3 + 1\right)$ when $t = 0$.

    Answer:
    $$\mathbf{v} = -2\mathbf{i}$$
    $$\mathbf{a} = -4\mathbf{i}$$

    Difficulty: 1   Section: 5

96. If $\mathbf{r}(t) = t^2\mathbf{i} + \sec t\,\mathbf{j}$, find the velocity and acceleration vectors.

    Answer:
    $$\mathbf{v} = 2t\mathbf{i} + \sec t \tan t\,\mathbf{j}$$
    $$\mathbf{a} = \left(\sec t \tan^2 t + \sec^3 t\right)\mathbf{j}$$

    Difficulty: 1   Section: 5

97. Write the integral for finding the length of the curve given by the equations $x = 4$, $y = t$, $z = t^2$, $0 \leq t \leq 2$.

    Answer: $\displaystyle\int_0^4 \sqrt{1 + 4t^2}\,dt$

    Difficulty: 1   Section: 5

98. Write the integral for finding the length of the curve given by the equations $x = e^t \sin t$, $y = e^t \cos t$, $z = 1$, $0 \leq t \leq 2$.

    Answer: $\displaystyle\int_0^2 \sqrt{2}e^t\,dt$

    Difficulty: 1   Section: 5

99. Write the integral for finding the length of the curve given by the equations $x = 1 + t^3$, $y = t$, $z = 1$, $0 \leq t \leq 1$.

    Answer: $\displaystyle\int_0^1 \sqrt{9t^4 + 1}\,dt$

    Difficulty: 1   Section: 5

100. Find the velocity and acceleration vectors for $x = \sin^2 t$, $y = \cos^2 t$, $z = t$.

    Answer:
    $$\mathbf{v} = (2\sin t \cos t)\mathbf{i} - (2\sin t \cos t)\mathbf{j} + \mathbf{k}$$
    $$\mathbf{a} = 2\left(\cos^2 t - \sin^2 t\right)\mathbf{i} + 2\left(\sin^2 t - \cos^2 t\right)\mathbf{j}$$

    Difficulty: 1   Section: 5

101. Find the velocity and acceleration vectors for $x = \dfrac{t+1}{t-1}$, $y = \ln\left(2t^3 + 1\right)$ and $z = t^3 - 5t$ when $t = 0$.

    Answer:
    $$\mathbf{v} = -2\mathbf{i} - 5\mathbf{k}$$

$\mathbf{a} = -4\mathbf{i}$

Difficulty: 1   Section: 5

102. If $\mathbf{r}(t) = t^2\,\mathbf{i} + \sec^2 t\,\mathbf{j} + \tan t\,\mathbf{k}$, find the velocity and acceleration vectors.

Answer:
$\mathbf{v} = 2t\,\mathbf{i} + \sec t \tan t\,\mathbf{j} + \sec^2 t\,\mathbf{k}$
$\mathbf{a} = 2\,\mathbf{i} + \left(\sec t \tan^2 t + \sec^3 t\right)\mathbf{j} + \left(2\sec^2 t \tan t\right)\mathbf{k}$

Difficulty: 1   Section: 5

103. Find the velocity and acceleration vectors if $\mathbf{r}(t) = \left(t^2 - t + 2\right)\mathbf{i} + (4\cos t - 3)\mathbf{j} + \left(e^{2t} - t\right)\mathbf{k}$.

Answer:
$\mathbf{v} = (2t - 1)\,\mathbf{i} + (-4\sin t)\,\mathbf{j} + \left(2e^{2t} - 1\right)\mathbf{k}$
$\mathbf{a} = 2\,\mathbf{i} - 4\cos t\,\mathbf{j} + 4e^{2t}\,\mathbf{k}$

Difficulty: 1   Section: 5

104. Write the parametric equations of the line through $(2, -1, 0)$ and $(-1, 2, 2)$.

Answer: $x = 2 + 3t;\ y = -1 - 3t;\ z = -2t$

Difficulty: 1   Section: 6

105. Write the symmetric equations of the line through $(2, -1, 0)$ and $(-1, 2, 2)$.

Answer: $\dfrac{x - 2}{3} = \dfrac{y + 1}{-3} = \dfrac{z}{2}$

Difficulty: 1   Section: 6

106. Write the parametric equations of the line through $(1, -4, -5)$ and $(4, -6, -1)$.

Answer: $x = 4 + 3t,\ y = -6 - 2t,\ z = -1 + 4t$

Difficulty: 1   Section: 6

107. Write the symmetric equations of the line through $(1, -4, -5)$ and $(4, -6, -1)$.

Answer: $\dfrac{x - 4}{3} = \dfrac{y + 6}{-2} = \dfrac{z + 1}{-2}$

Difficulty: 1   Section: 6

108. Write the symmetric equations of the line through $(0, 2, 3)$ and $(2, -4, -1)$.

Answer: $\dfrac{x - 2}{1} = \dfrac{y + 4}{-3} = \dfrac{z + 1}{-2}$

Difficulty: 1   Section: 6

109. Write the parametric equations of the line through $(0, 2, 3)$ and $(2, -4, -1)$.

Answer: $x = 2 + t,\ y = -4 - 3t,\ z = -1 - 2t$

Difficulty: 1   Section: 6

110. Write the parametric equations of the line through $(2, -3, 1)$ parallel to $3\mathbf{i} - \mathbf{j}$.

   Answer: $x = 2 + 3t$, $y = -3 - t$, $z = 1$

   Difficulty: 1   Section: 6

111. Write the symmetric equations of the line through $(2, -3, 1)$ parallel to $3i - j$.

   Answer: $\dfrac{x-2}{3} = \dfrac{y+3}{-1}$, $z = 1$

   Difficulty: 1   Section: 6

112. Write the parametric equations of the line through $(-2, 3, -1)$ and $(-1, 3, 4)$.

   Answer: $x = -1 + t$, $y = 3$, $z = 4 + 5t$

   Difficulty: 1   Section: 6

113. Write the symmetric equations of the line through $(-2, 3, -1)$ and $(-1, 3, 4)$.

   Answer: $\dfrac{x+1}{1} = \dfrac{z-4}{5}$, $y = 3$

   Difficulty: 1   Section: 6

114. Find a vector parallel to the line given by the equations $x = -3 - t$, $y = 2 + 5t$, $z = 3t$.

   Answer: any nonzero scalar multiple of $-\mathbf{i} + 5\mathbf{j} + 3\mathbf{k}$

   Difficulty: 1   Section: 6

115. Find a vector parallel to the line given by the equations $\dfrac{x+3}{-1} = \dfrac{y-2}{5} = \dfrac{z}{3}$.

   Answer: any nonzero scalar multiple of $-\mathbf{i} + 5\mathbf{j} + 3\mathbf{k}$

   Difficulty: 1   Section: 6

116. Find a vector parallel to the line given by the equations $x = 1$, $y = 9 - 2t$, $z = -5 - t$.

   Answer: Any scalar multiple of $2\mathbf{j} + \mathbf{k}$

   Difficulty: 1   Section: 6

117. Find a vector parallel to the line given by the equations $x = 1$, $\dfrac{y-9}{-2} = \dfrac{z+5}{-1}$.

   Answer: Any scalar multiple of $2\mathbf{j} + \mathbf{k}$

   Difficulty: 1   Section: 6

118. Find a point on the line given by the equations $\dfrac{x-3}{2} = \dfrac{y+1}{4} = \dfrac{z+2}{-1}$ such that the first coordinate is 4.

Answer: $\left(4, 1, -\dfrac{5}{2}\right)$

Difficulty: 2   Section: 6

119. Find a point on the line given by the equations $x = 3 + 2t$, $y = -1 + 4t$, $z = -2 - t$ such that the first coordinate is 4.

Answer: $\left(4, 1, -\dfrac{5}{2}\right)$

Difficulty: 2   Section: 6

120. Find an equation for the line tangent to the curve $x = \sin t$, $y = \cos t$, $z = 1$ at $(0, -1, 1)$.

Answer: $-\mathbf{i}$

Difficulty: 2   Section: 6

121. Find an equation for the line tangent to the curve $x = t^2$, $y = t^3$, $z = 1 - t$ at $(1, -1, 2)$.

Answer: $-2\mathbf{i} + 3\mathbf{j} - \mathbf{k}$

Difficulty: 2   Section: 6

122. Find a tangent vector for the curve $x = 3t \cos t$, $y = 3t \sin t$, $z = 4t$ in terms of t.

Answer: $(3\cos t - 3t \sin t)\mathbf{i} + (3 \sin t + 3t \cos t)\mathbf{j} + 4\mathbf{k}$

Difficulty: 2   Section: 6

123. Find a tangent vector for the curve $x = \sec t$, $y = \tan t$, $z = 3$, at $\left(\sqrt{2}, 1, 3\right)$.

Answer: $\sqrt{2}\mathbf{i} + 2\mathbf{j}$

Difficulty: 2   Section: 6

124. Find a unit vector tangent to the curve $x = 2\cos 6t$, $y = 2 \sin 6t$, $z = 5t$ at $\left(0, 2, \dfrac{5\pi}{12}\right)$.

Answer: $\dfrac{1}{13}(-12\mathbf{i} + 5\mathbf{k})$

Difficulty: 2   Section: 6

125. Find the curvature of the curve given by $x = t$, $y = \ln t$.

Answer: $\dfrac{t}{(1 + t^2)^{3/2}}$

Difficulty: 2   Section: 7

126. Find the curvature of the curve given by $x = \cos^3 t$, $y = \sin^3 t$.

Answer: $\dfrac{1}{3 \cos t \sin t}$

Difficulty: 2   Section: 7

127. Find the curvature of the curve given by $x = 3\cos 2t$, $y = \sin 2t$ at $(0, 1)$.

   Answer: $\dfrac{1}{9}$

   Difficulty: 2   Section: 7

128. Find the curvature of the curve given by $x = 4\cos 3t$, $y = \sin 3t$.

   Answer: $\dfrac{4}{\left(16\sin^2 5t + \cos^2 3t\right)^{3/2}}$

   Difficulty: 2   Section: 7

129. Find the curvature of the curve given by $x = 2\cos 5t$, $y = \sin 5t$.

   Answer: $\dfrac{2}{\left(4\sin^2 5t + \cos^2\right)}$

   Difficulty: 2   Section: 7

130. Find the curvature of the curve given in polar coordinates by $r = e^2\theta$.

   Answer: $\dfrac{1}{r\sqrt{5}}$

   Difficulty: 2   Section: 7

131. Find the curvature of the curve $y = \ln\sec x$.

   Answer: $\cos x$

   Difficulty: 2   Section: 7

132. Find the point on the parabola $y = x^2 + 2x$ where the curvature is a maximum.

   Answer: at the vertex $(-1, -1)$

   Difficulty: 2   Section: 7

133. Find the radius of curvature of the curve $y = x^3$ at $(1, 1)$.

   Answer: $\dfrac{5\sqrt{10}}{3}$

   Difficulty: 2   Section: 7

134. Find the radius of curvature of the curve $y = \dfrac{x^2}{2}$ at $\left(1, \dfrac{1}{2}\right)$.

   Answer: $2\sqrt{2}$

   Difficulty: 2   Section: 7

135. If $\mathbf{r}(t) = \sin^2 t \, \mathbf{i} + \cos^2 t \, \mathbf{j}$, find the velocity and acceleration vectors.

Answer:
$$\mathbf{v} = (2\sin t \cos t)\mathbf{i} + (-2\sin t \cos t)\mathbf{j} = \sin 2t \, \mathbf{i} - \sin 2t \, \mathbf{j}$$
$$\mathbf{a} = 2\left(\cos^2 t - \sin^2 t\right)\mathbf{i} + 2\left(\sin^2 t - \cos^2 t\right)\mathbf{j} = 2\cos 2t \, \mathbf{i} - 2\cos 2t \, \mathbf{j}$$

Difficulty: 1   Section: 7

136. If $\mathbf{r}(t) = \sin^2 t \, \mathbf{i} + \cos^2 t \, \mathbf{j}$, find the normal and tangential components of the acceleration.

Answer:
$$a_T = 2\sqrt{2}\cos 2t, \ a_N = 0$$

Difficulty: 1   Section: 7

137. Find the unit tangent and unit normal vectors to the curve given by $x = \ln t$, $y = 2t$.

Answer: $\mathbf{T} = \dfrac{\mathbf{i} + 2t\mathbf{j}}{\sqrt{1 + 4t^2}}$, $\mathbf{N} = \dfrac{-2t\mathbf{i} + \mathbf{j}}{\sqrt{1 + 4t^2}}$

Difficulty: 2   Section: 7

138. Find the unit tangent and unit normal vectors to the curve given by $x = 2t$, $y = \ln(\cos t)^2$.

Answer: $\mathbf{T} = \cos t \, \mathbf{i} - \sin t \, \mathbf{j}$, $\mathbf{N} = \sin t \, \mathbf{i} + \cos t \, \mathbf{j}$

Difficulty: 2   Section: 7

139. Find the unit tangent and unit normal vectors to the curve described by the function $\mathbf{r}(t) = t^2\mathbf{i} - 4t\mathbf{j}$.

Answer: $\mathbf{T} = \dfrac{t\mathbf{i} - 2\mathbf{j}}{\sqrt{4 + t^2}}$, $\mathbf{N} = \dfrac{2\mathbf{j} + t\mathbf{j}}{\sqrt{4 + t^2}}$

Difficulty: 2   Section: 7

140. Find the unit tangent and unit normal vectors to the curve described by the function $\mathbf{r}(t) = e^t \sin t \, \mathbf{i} + e^t \cos t \, \mathbf{j}$.

Answer: $\mathbf{T} = \dfrac{\cos t + \sin t}{\sqrt{2}}\mathbf{i} + \dfrac{\cos t - \sin t}{\sqrt{2}}\mathbf{j}$

$\mathbf{N} = \dfrac{\cos t + \sin t}{\sqrt{2}}\mathbf{i} + \dfrac{\cos t - \sin t}{\sqrt{2}}\mathbf{j}$

Difficulty: 2   Section: 7

141. Find the unit tangent and unit normal vectors to the curve described by the function $\mathbf{r}(t) = 3y \cos t \, \mathbf{i} + 3t \sin t \, \mathbf{j}$.

Answer: $\mathbf{T} = \dfrac{(\cos t - t\sin t)\mathbf{i} + (\sin t + t\cos t)\mathbf{j}}{\sqrt{1 + t^2}}$

$\mathbf{N} = \dfrac{(-\sin t - t\cos t)\mathbf{i} + (\cos t - t\sin t)\mathbf{j}}{\sqrt{1 + t^2}}$

Difficulty: 2   Section: 7

142. Given that $\mathbf{r}(t) = (3t+6)\mathbf{i} + (t^3)\mathbf{j} + \left(\dfrac{3\sqrt{2}}{2}t^2\right)\mathbf{k}$, find the curvature.

Answer: $\dfrac{\sqrt{2}}{3(1+t^2)^2}$

Difficulty: 2 Section: 7

143. Given that $\mathbf{r}(t) = (3t+6)\mathbf{i} + (t^3)\mathbf{j} + \left(\dfrac{3\sqrt{2}}{2}t^2\right)\mathbf{k}$, find the normal and tangential components of the acceleration.

Answer: $a_r = 6t,\ a_n = 3\sqrt{2}$

Difficulty: 2 Section: 7

144. Given that $\mathbf{r}(t) = (2\cos 6t)\mathbf{i} + (2\sin 6t)\mathbf{j} + 3j\,\mathbf{k}$, find the unit tangent vector and the principal normal vector.

Answer:
$$\mathbf{T} = \dfrac{1}{\sqrt{153}}(-12\sin 6t\,\mathbf{i} + 12\cos 6t\,\mathbf{j} + 3\mathbf{k})$$
$$\mathbf{N} = (-\cos 6t)\mathbf{i} + (-\sin 6t)\mathbf{j}$$

Difficulty: 2 Section: 7

145. Given that $\mathbf{r}(t) = (2\cos 6t)\mathbf{i} + (2\sin 6t)\mathbf{j} + 3t\,\mathbf{k}$, find the curvature.

Answer: $\dfrac{72}{153}$

Difficulty: 2 Section: 7

146. If $\mathbf{r}(t) = 2t^2\mathbf{i} + \dfrac{3}{2}t^2\mathbf{j} + 5t\,\mathbf{k}$, find the normal and tangential components of the acceleration.

Answer: $a_T = \dfrac{5t}{\sqrt{t^2+1}},\ a_N = \dfrac{5}{\sqrt{t^2+1}}$

Difficulty: 2 Section: 7

147. If $\mathbf{r}(t) = 2t^2\mathbf{i} + \dfrac{3}{2}t^2\mathbf{j} + 5t\,\mathbf{k}$, find the curvature.

Answer: $\dfrac{1}{5(1+t^2)^{3/2}}$

Difficulty: 2 Section: 7

148. If $\mathbf{r}(t) = 2t^2\mathbf{i} + \dfrac{3}{2}t^2\mathbf{j} + 5t\,\mathbf{k}$, find the unit tangent vector and the unit normal vector.

Answer:
$$\mathbf{T} = \dfrac{1}{\sqrt{t^2+1}}(4t\,\mathbf{i} + 3t\,\mathbf{j} + 5\mathbf{k})$$

$$\mathbf{N} = \frac{1}{5\sqrt{t^2+1}}(4\mathbf{i} + 3\mathbf{j} - 5t\,\mathbf{k})$$

Difficulty: 2   Section: 7

149. Classify the quadric surface whose equation is $\dfrac{x^2}{4} - \dfrac{y^2}{9} + z^2 = 1$

   Answer: hyperboloid of one sheet

   Difficulty: 1   Section: 8

150. Classify the quadric surface whose equation is $\dfrac{x^2}{4} + \dfrac{y^2}{9} + \dfrac{z^2}{16} = 1$

   Answer: ellipsoid

   Difficulty: 1   Section: 8

151. Classify the quadric surface whose equation is $\dfrac{x^2}{9} - \dfrac{y^2}{9} - z^2 = 1$

   Answer: hyperboloid of two sheets

   Difficulty: 1   Section: 8

152. Classify the quadric surface whose equation is $x^2 - y^2 + z^2 = 0$

   Answer: elliptic cone

   Difficulty: 1   Section: 8

153. Classify the quadric surface whose equation is $\dfrac{x^2}{4} + \dfrac{y^2}{9} = z$

   Answer: elliptic paraboloid

   Difficulty: 1   Section: 8

154. Write the equation of the surface $x^2 + z^2 = x$ in cylindrical coordinates.

   Answer: $r^2 \cos^2 \theta + z^2 = r \cos \theta$

   Difficulty: 1   Section: 9

155. Write the equation of the surface $z = x^2 + y^2 - y$ in cylindrical coordinates.

   Answer: $z = r(r - \sin \theta)$

   Difficulty: 1   Section: 9

156. Write the equation of the surface $x^2 + y^2 = 3x$ in cylindrical coordinates.

   Answer: $r = 3 \cos \theta$

   Difficulty: 1   Section: 9

157. Write the equation of the surface $x^2 + y^2 = 4y$ in cylindrical coordinates.

   Answer: $r = 4\sin\theta$

   Difficulty: 1   Section: 9

158. Write the equation $\rho = 2\sec\rho$ in rectangular coordinates.

   Answer: $z - 2$

   Difficulty: 1   Section: 9

159. Write the equation $z = 5$ in spherical coordinates.

   Answer: $\rho = 5\sec\theta$

   Difficulty: 1   Section: 9

160. Write the equation $x = 6$ in cylindrical coordinates.

   Answer: $r = 6\sec\theta$

   Difficulty: 1   Section: 9

161. Write the equation $x^2 - y^2 = 1$ in cylindrical coordinates.

   Answer: $r^2\cos^2\theta = 1$

   Difficulty: 1   Section: 9

12 Derivatives for Functions of Two or More Variables

1. What is the natural domain of the function $f(x,y) = \tan^{-1}\frac{y}{x}$?

   Answer: $\{(x,y) : x \neq 0\}$

   Difficulty: 1   Section: 1

2. What is the natural domain of the function $f(x,y) = \frac{2x-y}{x+3y}$?

   Answer: $\{(x,y) : x \neq -3y\}$

   Difficulty: 1   Section: 1

3. What is the natural domain of the function $f(x,y) = \frac{2x-y}{x-3y}$?

   Answer: $\{(x,y) : x \neq 3y\}$

   Difficulty: 1   Section: 1

4. If $f(x,y) = \tan^{-1}\frac{y}{x}$, what is $f(1,1)$?

   Answer: $\frac{\pi}{4}$

   Difficulty: 1   Section: 1

5. If $f(x,y) = \tan^{-1}\frac{y}{x}$, what is $f(1,0)$?

   Answer: 0

   Difficulty: 1   Section: 1

6. If $f(x,y) = \tan^{-1}\frac{y}{x}$, what is $f(\sqrt{3},1)$?

   Answer: $\frac{\pi}{6}$

   Difficulty: 1   Section: 1

7. If $f(x,y) = \tan^{-1}\frac{y}{x}$, what is $f(1,\sqrt{3})$?

   Answer: $\frac{\pi}{3}$

   Difficulty: 1   Section: 1

8. If $f(x,y) = \frac{2x-y}{x+3y}$, what is $f(c,0)$, $c \neq 0$?

   Answer: 2

   Difficulty: 1   Section: 1

9. If $f(x,y) = \dfrac{2x-y}{x+3y}$, what is $f(0,c)$, $c \neq 0$?

   Answer: $-\dfrac{1}{3}$

   Difficulty: 1   Section: 1

10. If $f(x,y) = -\dfrac{2x-y}{x+3y}$, what is $f(c,c)$, $c \neq 0$?

    Answer: $\dfrac{1}{4}$

    Difficulty: 1   Section: 1

11. If $f(x,y) = \ln(x^2+y^2)^{1/2}$, find $\dfrac{\partial f}{\partial x}$ and $\dfrac{\partial f}{\partial y}$.

    Answer: $\dfrac{\partial f}{\partial x} = x(x^2+y^2)^{-1}$, $\dfrac{\partial f}{\partial y} = y(x^2+y^2)^{-1}$

    Difficulty: 1   Section: 2

12. If $f(x,y) = \ln(x^2+y^2)^{1/2}$, find the second partial derivatives.

    Answer: $\dfrac{\partial^2 f}{\partial x^2} = \dfrac{y^2-x^2}{(x^2+y^2)^2}$, $\dfrac{\partial^2 f}{\partial y^2} = -\dfrac{\partial^2 f}{\partial x^2}$, $\dfrac{\partial^2 f}{\partial x \partial y} = \dfrac{\partial^2 f}{\partial y \partial x} = \dfrac{2xy}{(x^2+y^2)^2}$

    Difficulty: 1   Section: 2

13. If $f(x,y) = \tan^{-1}\dfrac{x}{y}$, find f_x and f_y.

    Answer: $f_x = \dfrac{y}{x^2+y^2}$, $f_y = \dfrac{-x}{x^2+y^2}$

    Difficulty: 1   Section: 2

14. If $f(x,y) = \tan^{-1}\dfrac{x}{y}$, find f_{xx}, f_{yy}, f_{xy}, and f_{yx}.

    Answer: $f_{xx} = \dfrac{-2xy}{(x^2+y^2)^2}$, $f_{yy} = -f_{xx}$, $f_{xy} = \dfrac{x^2-y^2}{(x^2+y^2)^2} = f_{yx}$

    Difficulty: 1   Section: 2

15. If $f(x,y) = \tan^{-1}\dfrac{y}{x}$, find f_x and f_y.

    Answer: $f_x = \dfrac{-y}{x^2+y^2}$, $f_y = \dfrac{x}{x^2+y^2}$

    Difficulty: 1   Section: 2

16. If $f(x,y) = \tan^{-1}\dfrac{y}{x}$, find f_{xx}, f_{yy}, f_{xy}, and f_{yx}.

Answer: $f_{xx} = \dfrac{2xy}{(x^2+y^2)^2}$, $f_{yy} = -f_{xx}$, $f_{xy} = \dfrac{y^2 - x^2}{(x^2+y^2)^2} = f_{yx}$

Difficulty: 1   Section: 2

17. If $f(x,y) = 2x^2y^3 + 17e^{xy}$, find f_x and f_v.

    Answer: $f_x = 4xy^3 + 17e^{xy}$, $f(x) = 6x^2y^2 + 17xe^{xy}$

    Difficulty: 1   Section: 2

18. If $f(x,y) = 2x^2y^3 + 17e^{xy}$, find f_{xx}, f_{yy}, f_{xy}, and f_{yx}.

    Answer: $f_{xx} = 4y^3 + 17y^2 e^{xy}$, $f_{yy} = 12x^2 y + 17x^2 e^{xy}$, $f_{xy} = 12xy^2 + 17e^{xy} + 17xye^{xy} = f_{yx}$

    Difficulty: 1   Section: 2

19. Let C be the curve formed the the intersection of the plane $y = -2$ and the surface $z = x^2 + 2xy - y^2$. Find the slope of the line tangent to C at $(1, -2, -7)$.

    Answer: -2

    Difficulty: 1   Section: 2

20. Let C be the curve formed the the intersection of the plane $x = 1$ and the surface $z = x^2 + 2xy - y^2$. Find the slope of the line tangent to C at $(1, -2, -7)$.

    Answer: 6

    Difficulty: 1   Section: 2

21. Let C be the curve formed the the intersection of the plane $x = 0$ and the surface $z = x^3 + y^3 - 9xy + 27$. Find the equations of the line tangent to C at $(0, -3, 0)$.

    Answer: $x = 0$, $z = 27t$, $y = -3 + t$

    Difficulty: 1   Section: 2

22. Let C be the curve formed the the intersection of the plane $y = -3$ and the surface $z = x^3 + y^3 - 9xy + 27$. Find the equations of the line tangent to C at $(0, -3, 0)$.

    Answer: $y = -3$, $z = 27t$, $x = t$

    Difficulty: 1   Section: 2

23. Find C so that $f(x,y) = x^3 + Cxy^2 - x^2 + y^2$ satisfies the equation $f_{xx} + f_{vv} = 0$.

    Answer: $C = 3$

    Difficulty: 2   Section: 2

24. Find C so that $f(x,y) = x^3 - 3xy^2 + Cx^2 + y^2$ satisfies the equation $f_{xx} + f_{vv} = 0$.

    Answer: $C = -1$

    Difficulty: 2   Section: 2.

25. Find C so that $f(x,y) = e^{-Cy} \sin 2x$ satisfies the equation $f_y = 4f_{xx}$.

   Answer: $C = 16$

   Difficulty: 2   Section: 2

26. Find C so that $f(x,y) = e^{-4y} \sin Cx$ satisfies the equation $f_y = f_{xx}$.

   Answer: $C = 2$ or -2

   Difficulty: 2   Section: 2

27. Find a so that $f(x,y) = \cos(ax + y)$ is a solution to the equation $f_{xx} = 9f_{vv}$.

   Answer: $a = 3$ or -3

   Difficulty: 2   Section: 2

28. Find a so that $f(x,y) = \cos(x + ay)$ is a solution to the equation $f_{yy} = 4f_{xx}$.

   Answer: $a = 2$ or -2

   Difficulty: 2   Section: 2

29. If $f(x,y,z) = x^3 y - \dfrac{xy}{z} + \tan^{-1} xy$, find f_x, f_y, and f_z.

   Answer:
   $$f_x = 3x^2 y - \frac{y}{z} + \frac{y}{1 + x^2 y^2}$$
   $$f_y = x^3 - \frac{x}{z} + \frac{x}{1 + x^2 y^2}, \; f_z = \frac{xy}{z^2}$$

   Difficulty: 1   Section: 2

30. If $f(x,y,z) = x^2 yz - \sin xyz + xe^{yz}$, find f_x, f_y, and f_z.

   Answer:
   $$f_x = 2xyz - yz \sin xyz + e^{yz}$$
   $$f_y = x^2 z - xz \sin xyz + xz \, e^{yz}$$
   $$f_z = x^2 y - xy \sin xyz + xy \, e^{yz}$$

   Difficulty: 1   Section: 2

31. Show that $\lim\limits_{(x,y)\to(0,0)} \dfrac{x - 7y}{x + y}$ doesn't exist.

   Answer: $f(0,y) = -1$ if $y \neq 0$ and $f(x,0) = 1$ if $x \neq 0$.

   Difficulty: 2   Section: 2

32. Show that $\lim\limits_{(x,y)\to(0,0)} \dfrac{x^2}{y}$ doesn't exist.

   Answer: along $x = 2\sqrt{y}$, $f(x,y) = 4$, along $x = \sqrt{y}$, $f(x,y) = 1$

   Difficulty: 2   Section: 3

33. Show that $\lim_{(x,y)\to(0,0)} \dfrac{x-y^2}{x+y^2}$ doesn't exist.

   Answer: along $x = 0$, $f(0, y) = -1$, along $y = 0$, $f(x, 0) = 1$ if $x \neq 0$.

   Difficulty: 2   Section: 3

34. Show that $\lim_{(x,y)\to(0,0)} \dfrac{2x-y}{x+3y}$ doesn't exist.

   Answer: along $x = 0$, $f(0, y) = -\dfrac{1}{3}$ if $y \neq 0$, along $y = 0$, $f(x, 0) = 2$ if $x \neq 0$.

   Difficulty: 2   Section: 3

35. Show that $\lim_{(x,y)\to(0,0)} \dfrac{x^2-2y}{x+3y}$ doesn't exist.

   Answer: along $x = 0$, $f(0, y) = -\dfrac{2}{3}$ if $y \neq 0$ and if $y = 0$, $f(x, 0) = x$ if $x \neq 0$.

   Difficulty: 2   Section: 3

36. Find the gradient ∇f for $f(x, y) = x^2 + 2xy - y^2$.

   Answer: $\nabla f(x, y) = (2x + 2y)\,\mathbf{i} + (2x - 2y)\,\mathbf{j}$

   Difficulty: 1   Section: 4

37. Find the gradient ∇f for $f(x, y) = x^3 + y^3 - 9xy + 27$.

   Answer: $\nabla f(x, y) = (3x^2 - 9y)\,\mathbf{i} + (3y^3 - 9x)\,\mathbf{j}$

   D

   Difficulty: 1   Section: 4

38. Find the gradient ∇f for $f(x, y) = \ln(x^2 + y^2)^{1/2}$.

   Answer: $\nabla f(x, y) = (x^2 + y^2)^{-1/2}(x\,\mathbf{i} + y\,\mathbf{j})$

   Difficulty: 1   Section: 4

39. Find the gradient ∇f for $f(x, y) = \tan^{-1}\dfrac{y}{x}$.

   Answer: $\nabla f(x, y) = \dfrac{-y}{x^2 + y^2}\,\mathbf{i} + \dfrac{x}{x^2 + y^2}\,\mathbf{j}$

   Difficulty: 1   Section: 4

40. Find the gradient ∇f for $f(x, y) = e^x \cos y$.

   Answer: $\nabla f(x, y) = (e^x \cos y)\,\mathbf{i} + (-e^x \sin y)\,\mathbf{j}$

   Difficulty: 1   Section: 4

41. Find the gradient ∇f for $f(x, y) = \sqrt{9 - x^2 - y^2}$.

Answer: $\nabla f(x,y) = (9 - x^2 - y^2)^{-1/2}(-x\,\mathbf{i} - y\,\mathbf{j})$

Difficulty: 1   Section: 4

42. Find the equation of the tangent plane for the surface $z = x^2 + 2xy - y^2$ at $(1, -2, -7)$.

    Answer: $z + 7 = -2(x - 1) + 6(y + 2)$

    Difficulty: 2   Section: 4

43. Find the equation of the tangent plane for the surface $z = x^3 + y^3 - 9xy + 27$ at $(0, -3, 0)$.

    Answer: $z = 27x + 27(y + 3)$

    Difficulty: 2   Section: 4

44. Find the equation of the tangent plane for the surface $z = \ln(x^2 + y^2)^{1/2}$ at $(1, 0, 0)$.

    Answer: $z = x - 1$

    Difficulty: 2   Section: 4

45. Find the equation of the tangent plane for the surface $z = \tan^{-1}\dfrac{y}{x}$ at $\left(1, 1, \dfrac{\pi}{4}\right)$.

    Answer: $z - \dfrac{\pi}{4} = -\dfrac{1}{2}(x - 1) + \dfrac{1}{2}(y - 1)$

    Difficulty: 2   Section: 4

46. Find the equation of the tangent plane for the surface $z = \sqrt{9 - x^2 - y^2}$ at $(1, -2, 2)$.

    Answer: $z - 2 = -\dfrac{1}{2}(x - 1) + (y + 2)$

    Difficulty: 2   Section: 4

47. Find the equation of the tangent plane for the surface $z = e^x \cos y$ at $(1, 0, e)$.

    Answer: $z - e = e(x - 1)$

    Difficulty: 2   Section: 4

48. Find the directional derivative of $f(x, y) = x^2 - 2y^2$ at $(3, 3)$ in the direction of $u = 2\mathbf{i} - \mathbf{j}$.

    Answer: $\dfrac{24}{\sqrt{5}}$

    Difficulty: 2   Section: 4

49. Find the directional derivative of $f(x, y) = x^2 - 2y^2$ at $(3, 3)$ in the direction of $y = 2\mathbf{i} + 3\mathbf{j}$.

    Answer: $-\dfrac{24}{\sqrt{13}}$

    Difficulty: 2   Section: 5

50. Find the directional derivative of $f(x,y) = x^2 - 2y^2$ at $(3,3)$ in the direction of $y = 2\mathbf{i} + \mathbf{j}$.

    Answer: 0

    Difficulty: 2   Section: 5

51. In what direction does $f(x,y) = x^2 - 2y^2$ increase most rapidly at $(3,3)$? What is the maximum rate of increase?

    Answer: $\dfrac{1}{\sqrt{5}}(\mathbf{i} - 2\mathbf{j})$; maximum rate is $\sqrt{180}$

    Difficulty: 2   Section: 5

52. In what direction does $f(x,y) = x^3y - x^2y^3$ increase most rapidly at $(2,-1)$? What is the maximum rate of increase?

    Answer: $\dfrac{-2\mathbf{i} - \mathbf{j}}{\sqrt{5}}(\mathbf{i} - 2\mathbf{j})$; maximum rate is $\sqrt{80}$

    Difficulty: 2   Section: 5

53. The function $f(x,y)$ at $(1,2)$ has a directional derivative equal to 2 in the direction towards $(2,2)$ and equal to -2 in the direction toward $(1,1)$. What is the directional derivative in the direction toward $(2,3)$?

    Answer: $\dfrac{4}{\sqrt{2}}$

    Difficulty: 3   Section: 5

54. The function $f(x,y)$ at $(1,2)$ has a directional derivative equal to 3 in the direction towards $(2,2)$ and equal to -1 in the direction toward $(1,1)$. What is the directional derivative in the direction toward $(2,3)$?

    Answer: $\dfrac{4}{\sqrt{2}}$

    Difficulty: 3   Section: 5

55. Find a unit vector $\mathbf{u}$ such that $D_{\mathbf{u}}f(2,-3) = 0$ where $f(x,y) = x^3y - x^2y$.

    Answer: $\dfrac{\mathbf{i} + 6\mathbf{j}}{\sqrt{37}}$

    Difficulty: 2   Section: 5

56. Find a unit vector $\mathbf{u}$ such that $D_{\mathbf{u}}f(2,-1) = 0$ where $f(x,y) = x^3y - x^2y^3$.

    Answer: $\dfrac{\mathbf{i} - 2\mathbf{j}}{\sqrt{5}}$

    Difficulty: 2   Section: 5

57. Find a unit vector $\mathbf{u}$ such that $D_{\mathbf{u}}f(1,-2) = 0$ where $f(x,y) = x^2 + 2xy - y^2$.

Answer: $\dfrac{3\mathbf{i}+\mathbf{j}}{\sqrt{10}}$

Difficulty: 2   Section: 5

58. Find the directional derivative of $f(x,y,z) = xy + yz - xz$ at $(1,-1,2)$ in the direction $\mathbf{i} - 2\mathbf{j} + 2\mathbf{k}$.

    Answer: $-\dfrac{1}{3}$

    Difficulty: 1   Section: 5

59. Find the directional derivative of $f(x,y,z) = xy + yz + xz$ at $(1,-1,2)$ in the direction $\mathbf{i} - 2\mathbf{j} + 2\mathbf{k}$.

    Answer: $-\dfrac{5}{3}$

    Difficulty: 1   Section: 5

60. Find the directional derivative of $f(x,y,z) = x^2 - 2y^2 + z^2$ at $(3,3,1)$ in the direction $2\mathbf{i} - \mathbf{j} + \mathbf{k}$.

    Answer: $\dfrac{26}{\sqrt{6}}$

    Difficulty: 1   Section: 5

61. In what direction does $f(x,y,z) = xy + yz + xz$ increase most rapidly at $(1,2,3)$? What is the rate of change in that direction?

    Answer: $5\mathbf{i} + 4\mathbf{j} + 3\mathbf{k}$, $\sqrt{50}$

    Difficulty: 2   Section: 5

62. Find the directional derivative of $f(x,y,z) = xy + 2x^2 - z^3$ at $(2,-3,-1)$ in the direction $\mathbf{i} - 2\mathbf{j} + 2\mathbf{k}$.

    Answer: $-\dfrac{5}{3}$

    Difficulty: 2   Section: 5

63. In what direction does $f(x,y,z) = xy + 2x^2 - z^3$ increase most rapidly at $(2,-3,-1)$?

    Answer: $\dfrac{5\mathbf{i} + 2\mathbf{j} - 3\mathbf{k}}{\sqrt{38}}$

    Difficulty: 2   Section: 5

64. Use the chain rule to find $\dfrac{df}{dt}$ if $f(x,y) = x^2 - 2xy + y^2$, $x = 2t + 1$ and $y = 3t - 7$.

    Answer: $2t - 16$

    Difficulty: 2   Section: 6

65. Use the chain rule to find $\dfrac{df}{dt}$ if $f(x,y) = x^2 - 2xy + y^2$, $x = 4 - 3t$ and $y = 2t + 1$.

Answer: $50t - 30$

Difficulty: 2   Section: 6

66. Use the chain rule to find $\dfrac{df}{dt}$ if $f(x,y) = x \ln y$ where $x = t^2$ and $y = e^t$.

    Answer: $3t^2$

    Difficulty: 2   Section: 6

67. Use the chain rule to find f_u if $f(x,y) = x^2 + 2y^2$ and $x = \sin u \cos v$, $y = \sin u \sin v$.

    Answer: $2 \cos u \sin u \left(\cos^2 v + 2 \sin^2 v\right)$

    Difficulty: 2   Section: 6

68. If $f(x,y) = x^3 + y^3 - 2xy + 18$, find $\dfrac{df}{dt}$ at $(2,1)$ if $\dfrac{dx}{dt} = -2$ and $\dfrac{dy}{dt} = 3$.

    Answer: -23

    Difficulty: 2   Section: 6

69. If $f(x,y) = x^3 + y^3 - 2xy + 18$, find $\dfrac{df}{dt}$ at $(2,1)$ if $\dfrac{dx}{dt} = -\dfrac{1}{2}$ and $\dfrac{dy}{dt} = -2$.

    Answer: -3

    Difficulty: 2   Section: 6

70. If z is defined implicitly as a function of x and y by $x^2 + 2xy + y^2 + z^2 = 6$, find $\dfrac{\partial z}{\partial x}$.

    Answer: $-\dfrac{x+y}{z}$

    Difficulty: 2   Section: 6

71. If z is defined implicitly as a function of x and y by $2x^2 + 3y^2 + 4x^2 = 9$, find $\dfrac{\partial z}{\partial y}$.

    Answer: $-\dfrac{3y}{4z}$

    Difficulty: 2   Section: 6

72. If z is defined implicitly as a function of x and y by $x^2 + 3xyz + 2xy^3 - z^3 = -15$, find $\dfrac{\partial z}{\partial x}$.

    Answer: $\dfrac{2x + 3yz}{3z^2 - 3xy}$

    Difficulty: 2   Section: 6

73. Determine the tangent plane and normal line to the surface $x^2 + 2xy - y^2 + z^2 = 6$ at $(1, 1, -2)$.

    Answer: $4(x-1) - 4(z+2) = 0$; $\dfrac{x-1}{4} = \dfrac{z+2}{-4}$, $y = 1$

Difficulty: 1   Section: 7

74. Determine the tangent plane and normal line to the surface $2x^2 + 3y^2 + 4z^2 = 9$ at $(1,1,1)$.

    Answer: $4(x-1) + 6(y-1) + 8(z-1) = 0$; $\dfrac{x-1}{4} = \dfrac{y-1}{6} = \dfrac{z-1}{8}$

    Difficulty: 1   Section: 7

75. Determine the tangent plane and normal line to the surface $x^3 + 3xyz + 2y^3 - z^3 = -15$ at $(1,-1,2)$.

    Answer: $-3(x-1) + 12(x+1) - 15(z-2) = 0$; $\dfrac{x-1}{-3} = \dfrac{y+1}{12} = \dfrac{z-2}{-15}$

    Difficulty: 1   Section: 7

76. Determine the tangent plane and normal line to the surface $xy + yz + xz - 1 = 0$ at $(3,-1,2)$.

    Answer: $(x-3) + 5(y+1) + 2(z-2) = 0$; $\dfrac{x-3}{1} = \dfrac{v+1}{5} = \dfrac{z-2}{2}$

    Difficulty: 1   Section: 7

77. Find the points on the surface $5x^2 - 6xy + 5y^2 - 8x + 8y + z^2 - 3z = 0$ has a tangent plane that is parallel to the x,y plane.

    Answer: $\left(\dfrac{1}{2}, -\dfrac{1}{2}, 4\right) \left(\dfrac{1}{2}, -\dfrac{1}{2}, -1\right)$

    Difficulty: 2   Section: 7

78. Find all relative maxima, relative minima and saddle points for $f(x,y) = x^3 + y^3 - 9xy + 27$.

    Answer: $(0,0)$ is a saddle point, $(3,3)$ is a relative minimum

    Difficulty: 1   Section: 8

79. Find all relative maxima, relative minima and saddle points for $f(x,y) = x^3 + y^3 + 3xy + 18$.

    Answer: $(0,0)$ is a saddle point, $(-1,-1)$ is a relative maximum

    Difficulty: 1   Section: 8

80. Find all relative maxima, relative minima and saddle points for $f(x,y) = x^3 + y^3 - 3xy + 9$.

    Answer: $(0,0)$ is a saddle point, $(1,1)$ is a relative minimum

    Difficulty: 1   Section: 8

81. Find all relative maxima, relative minima and saddle points for $f(x,y) = x^2 - 4xy + y^3 + 4y$.

    Answer: $(4,2)$ is a relative minimum, $\left(\dfrac{4}{3}, \dfrac{2}{3}\right)$ is a saddle point

    Difficulty: 1   Section: 8

82. Find all relative maxima, relative minima and saddle points for $f(x,y) = x^2 + 2y^2 - 4x + 4y - 3$.

    Answer: $(2, -1)$ is a relative minimum

    Difficulty: 1   Section: 8

83. Find all relative maxima, relative minima and saddle points for $f(x,y) = x^3 + y^2 - 6x^2 + y - 1$.

    Answer: $\left(4, -\dfrac{1}{2}\right)$ is a relative minimum, $\left(0, -\dfrac{1}{2}\right)$ is a saddle point

    Difficulty: 1   Section: 8

84. Find all relative maxima, relative minima and saddle points for $f(x,y) = \dfrac{1}{x} - \dfrac{27}{y} + xy$.

    Answer: $\left(-\dfrac{1}{3}, 9\right)$ is a relative maximum

    Difficulty: 1   Section: 8

85. Find all relative maxima, relative minima and saddle points for $f(x,y) = \dfrac{1}{x} - \dfrac{1}{y} + xy$.

    Answer: $(-1, 1)$ is a relative maximum

    Difficulty: 1   Section: 8

86. Find all relative maxima, relative minima and saddle points for $f(x,y) = x^2 - 3xy + 5x - 2y + 2y^2 + 8$.

    Answer: $(14, 11)$ is a saddle point

    Difficulty: 1   Section: 8

87. Find all relative maxima, relative minima and saddle points for $f(x,y) = x^2 - 3xy - x + 2y + 2y^2 + 9$.

    Answer: $(2, 1)$ is a saddle point

    Difficulty: 1   Section: 8

88. Find all relative maxima, relative minima and saddle points for $f(x,y) = x^2 - 3xy + 4x - 5y + 2y^2 + 10$.

    Answer: $(1, 2)$ is a saddle point

    Difficulty: 1   Section: 8

89. Find the maximum value of $f(x,y) = y^2 - 3x^2 + 2x$ if $x^2 + y^2 = 9$.

    Answer: $9\dfrac{1}{4}$ at $x = \dfrac{1}{4}, y = \dfrac{\sqrt{143}}{4}$

    Difficulty: 2   Section: 8

90. Find the distance from the point $(-6, 2, 3)$ to the plane $4x - 5y + 8z = 7$.

    Answer: $\dfrac{17}{\sqrt{105}}$

    Difficulty: 1   Section: 8

91. Find the distance from the point $(4, -3, 1)$ to the plane $2x + 3y - 6z = -14$.

    Answer: $\dfrac{17}{\sqrt{105}}$

    Difficulty: 1   Section: 8

92. Find the largest and smallest values of $f(x, y) = 2x^2 + y^2 - y$ on the circle $x^2 + y^2 = 1$.

    Answer: $f(0, 1) = 0$ and $f\left(\pm\dfrac{\sqrt{3}}{2}, -\dfrac{1}{2}\right) = \dfrac{5}{4}$

    Difficulty: 2   Section: 8

93. Find the largest and smallest value of $f(x, y) = x^2 + y^2$ on the parabola $x^2 + y = 1$.

    Answer: $f\left(\pm\dfrac{\sqrt{2}}{2}, \dfrac{1}{2}\right) = \dfrac{3}{4}$

    Difficulty: 2   Section: 9

94. Find the largest and smallest value of $f(x, y) = x - x^2 - y^2$ if $x^2 + y^2 = 4$.

    Answer: $f(2, 0) = -2$, $f(-2, 0) = -6$

    Difficulty: 2   Section: 9

95. Find the largest and smallest value of $f(x, y) = x - x^2 - y^2$ if $x^2 + y^2 = 1$.

    Answer: $f(1, 0) = 0$, $f(-1, 0) = -2$

    Difficulty: 2   Section: 9

96. Find the largest and smallest value of $f(x, y) = x^2 + y^2$ on the parabola $x^2 + y = \dfrac{3}{2}$.

    Answer: $f\left(\pm 1, \dfrac{1}{2}\right) = \dfrac{5}{4}$

    Difficulty: 2   Section: 9

97. Find the largest and smallest value of $f(x, y) = x^2 + y$ if $x^2 + y^2 = 4$.

    Answer: $f(0, -2) = -2$ and $f\left(\pm\dfrac{\sqrt{15}}{2}, \dfrac{1}{2}\right) = \dfrac{17}{4}$

    Difficulty: 2   Section: 9

13   Multiple Integrals

1. Compute $\iint_D 2\, dA$ where D is given by $1 \leq x \leq 3$, $-1 \leq y \leq 2$.

   Answer: 12

   Difficulty: 1   Section: 1

2. Compute $\iint_D 3\, dA$ where D is given by $1 \leq x \leq 3$, $2 \leq y \leq 6$.

   Answer: 24

   Difficulty: 1   Section: 1

3. Compute $\iint_D 5\, dA$ where D is bounded by $x^2 + y^2 = 2$.

   Answer: 10π

   Difficulty: 1   Section: 1

4. Compute $\iint_D 4\, dA$ where D is bounded by $x^2 + y^2 = 1$.

   Answer: 4π

   Difficulty: 1   Section: 1

5. Compute $\iint_D 3\, dA$ where D is bounded by $(x-1)^2 + (y+2)^2 = 3$.

   Answer: 9π

   Difficulty: 1   Section: 1

6. Compute $\iint_D 2\, dA$ where D is bounded by $x^2 + (y-7)^2 = 1$.

   Answer: 2π

   Difficulty: 1   Section: 1

7. Evaluate the integral $\int_0^1 \int_0^2 \left(x^2 + y^2\right) dy\, dx$.

   Answer: $\dfrac{10}{3}$

   Difficulty: 1   Section: 2

8. Evaluate the integral $\int_0^1 \int_0^3 (x^2 + y^2)\, dy\, dx$.

   Answer: 10

   Difficulty: 1   Section: 2

9. Evaluate the integral $\int_0^2 \int_0^1 x^2 y\, dx\, dy$.

   Answer: $\dfrac{2}{3}$

   Difficulty: 1   Section: 2

10. Evaluate the integral $\int_0^1 \int_0^2 x^2 y\, dx\, dy$.

    Answer: $\dfrac{4}{3}$

    Difficulty: 1   Section: 2

11. Evaluate the integral $\int_1^4 \int_0^2 (x + y^2)\, dx\, dy$.

    Answer: 48

    Difficulty: 1   Section: 2

12. Find the volume of the solid above the rectangle $0 \le x \le 1$, $0 \le y \le 2$ and below the surface $z = x^2 + y^2$.

    Answer: $\dfrac{10}{3}$

    Difficulty: 1   Section: 2

13. Find the volume of the solid above the rectangle $0 \le x \le 1$, $0 \le y \le 3$ and below the surface $z = x^2 + y^2$.

    Answer: 10

    Difficulty: 1   Section: 2

14. Find the volume of the solid above the rectangle $0 \le x \le 1$, $0 \le y \le 2$ and below the surface $z = x^2 y$.

    Answer: $\dfrac{2}{3}$

    Difficulty: 1   Section: 2

15. Find the volume of the solid above the rectangle $0 \le x \le 2$, $0 \le y \le 1$ and below the surface $z = x^2 y$.

Answer: $\dfrac{4}{3}$

Difficulty: 1   Section: 2

16. Find the volume of the solid above the rectangle $0 \leq x \leq 2$, $1 \leq y \leq 4$ and below the surface $z = x + y^2$.

   Answer: 48

   Difficulty: 1   Section: 2

17. Compute $\iint_D 4\,dA$ where D is bounded by $y = x^2$, the x-axis, and the line $x = 1$.

   Answer: $\dfrac{4}{3}$

   Difficulty: 1   Section: 3

18. Compute $\iint_D 3\,dA$ where D is bounded by $y = x^2$, the x-axis, and the line $x = 2$.

   Answer: 8

   Difficulty: 1   Section: 3

19. Compute $\iint_D 3\,dA$ where D is bounded by $y = 1 - x^2$ and the x-axis.

   Answer: 4

   Difficulty: 1   Section: 3

20. Write a double integral to express the volume of the solid with base D where D is the region bounded by $x^2 + 2y^2 = 4$ and the solid extends from each point (x, y) in the base to the surface $z = 2x^2 + y^2$.

   Answer: $\iint_D (2x^2 + y^2)\,dA$

   Difficulty: 1   Section: 3

21. Write a double integral to express the volume of the solid with base D where D is the region bounded by $x^2 + 5y^2 = 24$ and the solid extends from each point (x, y) in the base to the surface $z = 3x^2 + 5y^2$.

   Answer: $\iint_D (3x^2 + 5y^2)\,dA$

   Difficulty: 1   Section: 3

22. Write a double integral to express the volume of the solid with base D where D is the region

bounded by x^2, $y = 0$ $x = 1$, and the solid extends from each point (x, y) in the base to the surface $z = 4xy$.

Answer: $\iint_D 4xy \, dA$

Difficulty: 1   Section: 3

23. Evaluate the integral $\iint_D xy \, dA$ where D is the region bounded by $y = x^2$, the x-axis and the line $x = 2$.

Answer: $\dfrac{16}{3}$

Difficulty: 1   Section: 3

24. Evaluate $\iint_D \sin \pi x^2 \, dA$ where D is the triangle with vertices $(0, 0)$, $(1, 0)$, and $(1, 2)$.

Answer: $\dfrac{2}{\pi}$

Difficulty: 2   Section: 3

25. Evaluate $\iint_D x \, dA$ where D is the region bounded by $y = x^2$ and $y = 2x$.

Answer: $\dfrac{4}{3}$

Difficulty: 1   Section: 3

26. Evaluate $\iint_D x \, dA$ where D is the region in the first quadrant bounded by $y = 4 - x^2$, $x = 0$. and $y = 0$.

Answer: 4

Difficulty: 1   Section: 3

27. Evaluate $\iint_D y \sin xy \, dA$ where D is the rectangle given by $1 \le x \le 2$, $0 \le y \le \pi$.

Answer: 0

Difficulty: 1   Section: 3

28. Evaluate $\iint_D (x^2 + 2xy) \, dA$ where D is the region between $y = x^2$ and $y = x^3$.

Answer: $\dfrac{3}{40}$

Difficulty: 1   Section: 3

29. Evaluate $\iint_D 3\,dA$ where D is the region bounded by $x = 0$, $y = e^x$ and $y = 2$.

    Answer: $3(\ln 4 - 1)$

    Difficulty: 1   Section: 3

30. Find the volume of the solid above the region bounded by $y = x^2$, the x-axis and the line $x = 2$, and below the surface $z = xy$.

    Answer: $\dfrac{16}{3}$

    Difficulty: 1   Section: 3

31. Find the volume of the solid above the triangle with vertices $(0, 0)$, $(1, 0)$, and $(1, 2)$ below the surface $z = \sin \pi x^2$.

    Answer: $\dfrac{2}{\pi}$

    Difficulty: 1   Section: 3

32. Find the volume of the solid above the region bounded by $y = x^2$ and $y = 2x$ below the surface $z = x$.

    Answer: $\dfrac{4}{3}$

    Difficulty: 1   Section: 3

33. Find the volume of the solid above the region in the first quadrant bounded by $y = 4 - x^2$, $x = 0$, and $y = 0$, below the surface $z = x$.

    Answer: 4

    Difficulty: 1   Section: 3

34. Find the volume of the solid if the first octant bounded by the coordinate planes and the plane $2x + 3y + 6x = 24$.

    Answer: 64

    Difficulty: 2   Section: 3

35. Find the volume of the solid in the first octant bounded by the coordinate planes and the plane $x + 2y + 3x = 8$.

    Answer: $\dfrac{128}{9}$

    Difficulty: 2   Section: 3

36. Find the volume of the solid above the region bounded by $y = x^2$ and $y = x^3$ and below the

E222   Chapter 13   Exam Questions

surface $z = x^2 + 2xy$.

Answer: $\dfrac{3}{40}$

Difficulty: 1   Section: 3

37. Reverse the order of integration: $\displaystyle\int_0^1 \int_0^{\sqrt{1-x}} f(x,y)\, dy\, dx$.

Answer: $\displaystyle\int_0^1 \int_0^{1-y^2} f(x,y)\, dy\, dx$

Difficulty: 2   Section: 3

38. Reverse the order of integration: $\displaystyle\int_0^1 \int_y^{\sqrt{y}} f(x,y)\, dx\, dy$.

Answer: $\displaystyle\int_0^1 \int_{x^2}^{x} f(x,y)\, dy\, dx$

Difficulty: 2   Section: 3

39. Reverse the order of integration: $\displaystyle\int_0^2 \int_0^{4-x^2} f(x,y)\, dy\, dx$.

Answer: $\displaystyle\int_0^4 \int_0^{\sqrt{4-y}} f(x,y)\, dx\, dy$

Difficulty: 2   Section: 3

40. Reverse the order of integration: $\displaystyle\int_0^{\sqrt{\pi/2}} \int_{y^2}^{\pi/2} f(x,y)\, dx\, dy$.

Answer: $\displaystyle\int_0^{\pi/2} \int_0^{\sqrt{x}} f(x,y)\, dy\, dx$

Difficulty: 2   Section: 3

41. Reverse the order of integration: $\displaystyle\int_0^4 \int_{y/2}^{\sqrt{y}} f(x,y)\, dx\, dy$.

Answer: $\displaystyle\int_0^2 \int_{x^2}^{2x} f(x,y)\, dy\, dx$

Difficulty: 2   Section: 3

42. Reverse the order of integration: $\displaystyle\int_1^2 \int_0^{\ln y} f(x,y)\, dx\, dy$.

Answer: $\displaystyle\int_0^{\ln 2} \int_{e^x}^{2} f(x,y)\, dy\, dx$

Difficulty: 2   Section: 3

43. Reverse the order of integration: $\int_{-1}^{2} \int_{x^2}^{x+2} f(x,y) \, dy \, dx$.

   Answer: $\int_{0}^{1} \int_{-\sqrt{y}}^{\sqrt{y}} f(x,y) \, dx \, dy + \int_{1}^{4} \int_{y-2}^{\sqrt{y}} f(x,y) \, dx \, dy$

   Difficulty: 2   Section: 3

44. Reverse the order of integration: $\int_{0}^{2} \int_{x^2}^{4x-x^2} f(x,y) \, dy \, dx$.

   Answer: $\int_{0}^{4} \int_{2-\sqrt{4-y}}^{\sqrt{y}} f(x,y) \, dx \, dy$

   Difficulty: 2   Section: 3

45. Evaluate $\iint_R \sin\theta \, dA$ over the region R given by $0 \le \theta \le \dfrac{\pi}{2}, \; 0 \le r \le \sqrt{\cos\theta}$.

   Answer: $\dfrac{1}{4}$

   Difficulty: 1   Section: 4

46. Find the volume of the solid above the region bounded by $r = 2\cos\theta$ and below the surface $z = 1 - y$.

   Answer: π

   Difficulty: 1   Section: 4

47. Find the area inside the circle $r = 3\cos\theta$ and outside the cardioid $r = 1 + \cos\theta$.

   Answer: π

   Difficulty: 1   Section: 4

48. Find the area inside the circle $r = 3\sin\theta$ and outside the cardioid $r = 1 + \sin\theta$.

   Answer: π

   Difficulty: 1   Section: 4

49. Find the volume of the solid above the region bounded by $x^2 + y^2 = 2y$ and below the surface $z = x^2 + y^2$.

   Answer: $\dfrac{3\pi}{2}$

   Difficulty: 2   Section: 4

50. Use polar coordinates to evaluate $\int_{0}^{1} \int_{0}^{\sqrt{1-x^2}} e^{-(x^2+y^2)} \, dy \, dx$.

Answer: $\dfrac{\pi}{4}\left(1 - \dfrac{1}{e}\right)$

Difficulty: 2   Section: 4

51. Use polar coordinates to evaluate $\displaystyle\int_0^2 \int_y^{\sqrt{4-y^2}} \dfrac{1}{1+x^2+y^2}\, dx\, dy$.

Answer: $\dfrac{\pi \ln 5}{8}$

Difficulty: 2   Section: 4

52. Find the volume of the solid above the region bounded by $x^2 + y^2 = 2x$ and below the surface $z = x^2 + y^2$.

Answer: $\dfrac{3\pi}{2}$

Difficulty: 2   Section: 4

53. Find the mass of a thin lamina bounded by $y = 0$, $y = \sin x$, $0 \le x \le \pi$ if the density at (x, y) is $2y$.

Answer: $\dfrac{\pi}{2}$

Difficulty: 2   Section: 5

54. Find the mass of a thin lamina bounded by $y = x^2$, $y = 0$, and $x = 1$ if the density at (x, y) is $x + y$.

Answer: $\dfrac{7}{20}$

Difficulty: 2   Section: 5

55. Find the mass of a thin lamina bounded by $x = 1 - y^2$, $x = 0$ and $y = 0$ if the density at (x, y) is x.

Answer: $\dfrac{4}{15}$

Difficulty: 2   Section: 5

56. Find the mass of a thin lamina bounded by $y = 0$, $y = x^2$, and $x = 2$ if the density at (x, y) is xy.

Answer: $\dfrac{16}{3}$

Difficulty: 2   Section: 5

57. Find the center of mass of the spherical shell between $x^2 + y^2 + z^2 = 1$ and $x^2 + y^2 + z^2 = 4$, $z \ge 0$. Assume $\delta(x, y) = 1$.

Answer: $\left(0, 0, \dfrac{45}{56}\right)$

Difficulty: 2   Section: 5

58. Find the movement of inertia about the x-axis of the region bounded by $y = x^2$, $y = 0$, and $x = 2$. Assume density factor of 1.

    Answer: $\dfrac{128}{21}$

    Difficulty: 2   Section: 5

59. Find the surface area of the portion of the surface $z = \sqrt{16 - x^2}$ inside the cylinder $x^2 + y^2 = 16$.

    Answer: 64

    Difficulty: 2   Section: 6

60. Find the surface area of the region formed by the intersection of the cylinders $x^2 + y^2 = 1$ and $x^2 + z^2 = 4$.

    Answer: 16

    Difficulty: 2   Section: 6

61. Find the surface area of the region formed by the intersection of the cylinders $x^2 + y^2 = 4$ and $x^2 + z^2 = 4$.

    Answer: 64

    Difficulty: 2   Section: 6

62. Find the surface area of the region formed by the intersection of the cylinders $9x^2 + 9y^2 = 1$ and $9x^2 + 9z^2 = 1$.

    Answer: $\dfrac{16}{9}$

    Difficulty: 2   Section: 6

63. Find the surface area of the portion of the surface of the sphere $x^2 + y^2 + z^2 = 4$ cut by the cylinder $x^2 + y^2 = 2y$.

    Answer: $8(\pi - 2)$

    Difficulty: 2   Section: 6

64. Find the surface area of the portion of the surface of the sphere $x^2 + y^2 + z^2 = 1$ cut by the cylinder $x^2 + y^2 = y$.

    Answer: $2(\pi - 2)$

    Difficulty: 2   Section: 6

65. Find the surface area of the portion of the cylinder $x^2 + y^2 = y$ inside the sphere $x^2 + y^2 + z^2 = 1$.

    Answer: 4

Difficulty: 2   Section: 6

66. Find the surface area of the portion of the cylinder $x^2+y^2=2y$ inside the sphere $x^2+y^2+z^2=4$.

   Answer: 16

   Difficulty: 2   Section: 6

67. Compute $\iiint_V (xz+3z)\,dV$ where V is the region in the first octant bounded by $x^2+z^2=9$ and $x+y=3$.

   Answer: $\dfrac{324}{5}$

   Difficulty: 2   Section: 7

68. Compute $\iiint_V (xz+2z)\,dV$ where V is the region in the first octant bounded by $x^2+z^2=4$ and $x+y=2$.

   Answer: $\dfrac{128}{5}$

   Difficulty: 2   Section: 7

69. Compute $\iiint_V (xz+z)\,dV$ where V is the region in the first octant bounded by $x^2+z^2=1$ and $x+y=1$.

   Answer: $\dfrac{4}{15}$

   Difficulty: 2   Section: 7

70. Compute $\iiint_V z^2\,dV$ where V is the region in the first octant bounded by $z=2$, $y=2z$, and $x=y+z$.

   Answer: $\dfrac{128}{5}$

   Difficulty: 2   Section: 7

71. Compute $\displaystyle\int_0^1 \int_{2x}^{x+y} (x+y+z)\,dz\,dy\,dx$.

   Answer: $\dfrac{13}{8}$

   Difficulty: 2   Section: 7

72. Evaluate $\displaystyle\int_{-1}^1 \int_{3x^2}^{4-x^2} \int_0^{6-z} dy\,dz\,dx$.

Answer: $\dfrac{304}{15}$

Difficulty: 2   Section: 7

73. Evaluate $\displaystyle\int_0^1 \int_0^x \int_0^{x-y} x\, dz\, dy\, dx$.

Answer: $\dfrac{1}{8}$

Difficulty: 1   Section: 7

74. Find the volume of the region in the first octant bounded by $x^2 + y^2 = 4$, the plane $y + z = 2$ and the plane $z = 0$.

Answer: $\dfrac{2}{3}(3\pi - 4)$

Difficulty: 2   Section: 7

75. Find the volume of the region in the first octant bounded by $x^2 + y^2 = 1$, the plane $y + z = 1$ and the plane $z = 0$.

Answer: $\dfrac{1}{12}(3\pi - 4)$

Difficulty: 2   Section: 7

76. Find the volume of the region formed by the intersection of the cylinders $x^2 + y^2 = 4$ and $x^2 + z^2 = 4$.

Answer: $\dfrac{138}{3}$

Difficulty: 2   Section: 7

77. Find the volume of the region formed by the intersection of the cylinders $x^2 + y^2 = 1$ and $x^2 + z^2 = 1$.

Answer: $\dfrac{16}{3}$

Difficulty: 2   Section: 7

78. Use cylindrical coordinates to find the volume of the region in the first octant bounded by the cylinder $x^2 + y^2 = 4$ and the plane $z = 1 + 2x + 3y$.

Answer: $\pi + \dfrac{40}{3}$

Difficulty: 2   Section: 8

79. Use cylindrical coordinates to find the volume of the region in the first octant bounded by the cylinder $x^2 + y^2 = 4$ and the plane $z = 1 + 3x + 3y$.

Answer: $\pi + 16$

Difficulty: 2   Section: 8

80. Use cylindrical coordinates to find the volume of the region in the first octant bounded by the cylinder $x^2 + y^2 = 4$ and the plane $z = 1 + x + 2y$.

    Answer: $\pi + 8$

    Difficulty: 2   Section: 8

81. Use cylindrical coordinates to find the volume of the region in the first octant bounded by the cylinder $x^2 + y^2 = 9$ and the plane $zx^2 + z^2 = 9$.

    Answer: 18

    Difficulty: 2   Section: 8

82. Use cylindrical coordinates to find the volume of the region in the first octant bounded by $x^2 + y^2 = 16$ and $x^2 + z^2 = 16$.

    Answer: $\dfrac{128}{3}$

    Difficulty: 2   Section: 8

83. Use cylindrical coordinates to find the volume of the region in the first octant bounded by the cone $z = r$ and the cylinder $r = 2\sin\theta$.

    Answer: $\dfrac{16}{9}$

    Difficulty: 2   Section: 8

84. Use cylindrical coordinates to find the volume of the region in the first octant bounded by the cone $z = 2r$ and the cylinder $r = 3\sin\theta$.

    Answer: 12

    Difficulty: 2   Section: 8

85. Use cylindrical coordinates to find the volume of the region inside the paraboloid $z = 9 - x^2 - y^2$, outside the cylinder $x^2 + y^2 = 4$, and above the plane $z = 0$.

    Answer: $\dfrac{25\pi}{2}$

    Difficulty: 2   Section: 8

86. Find the volume of the region within the cylinder $z = 2\sin\theta$ bounded above by the paraboloid $z = r^2$ and below by the plane $z = 0$.

    Answer: $\dfrac{3\pi}{2}$

    Difficulty: 2   Section: 8

87. Find the volume of the solid bounded by the plane $z = 0$, the plane $z = 1 - y$, and the cylinder

$r = 2\cos\theta$.

Answer: π

Difficulty: 2   Section: 8

88. The plane $z = 1$ divides the solid sphere $x^2 + y^2 + z^2 \leq 2$ into two parts. Find the volume of the smaller part.

Answer: $\dfrac{\pi}{3}\left(4\sqrt{2} - 5\right)$

Difficulty: 2   Section: 8

89. Evaluate $\iiint\limits_V z^2\, dV$ where V is the region described by $x^2 + y^2 + z^2 \leq 9$, $x \geq 0$, $y \geq 0$, and $z \geq 0$.

Answer: $\dfrac{81\pi}{10}$

Difficulty: 2   Section: 8

90. Find the volume of the region above the cone $z = \sqrt{x^2 + y^2}$ and below the sphere $x^2 + y^2 + z^2 = z$.

Answer: $\dfrac{\pi}{8}$

Difficulty: 2   Section: 8

91. Find the volume of the region above the plane $z = 3$ and below the sphere $x^2 + y^2 + z^2 = 18$.

Answer: $9\pi\left(4\sqrt{2} - 5\right)$

Difficulty: 2   Section: 8

92. Find the volume of the region above the plane $z = 2$ and below the sphere $x^2 + y^2 + z^2 = 8$.

Answer: $\dfrac{8\pi}{3}\left(4\sqrt{2} - 5\right)$

Difficulty: 2   Section: 8

93. Find the volume of the region above the plane $z = 4$ and below the sphere $x^2 + y^2 + z^2 = 32$.

Answer: $\dfrac{64\pi}{3}\left(4\sqrt{2} - 5\right)$

Difficulty: 2   Section: 8

94. Find the Jacobian for the transformation $x = r\sin t$, $y = r\cos t$.

Answer: $-r$

Difficulty: 1   Section: 9

95. Evaluate the double integral $\iint_R \sin(x-y)\cos(x+y)\,dA$, where R is the triangle with vertices $(0,0)$, $(\pi,-\pi)$ and (π,π).

Answer: $-\dfrac{\pi}{2}$

Difficulty: 2   Section: 9

14 Vector Calculus

1. If $\mathbf{F} = x^2\,\mathbf{i} + 2xy^2z\,\mathbf{j} + \cos z\,\mathbf{k}$, find $\nabla \cdot \mathbf{F}$.

   Answer: $2x + 4xyz - \sin z$

   Difficulty: 1   Section: 1

2. If $\mathbf{F} = x^2\,\mathbf{i} + 2xy^2z\,\mathbf{j} + \cos z\,\mathbf{k}$, find $\nabla \times \mathbf{F}$.

   Answer: $-2x^2y\,\mathbf{i} + 2y^2z\,\mathbf{k}$

   Difficulty: 1   Section: 1

3. If $\mathbf{F} = (x^2 - y^2)\,\mathbf{i} + 2xyz\,\mathbf{j} + z^2\,\mathbf{k}$, find $\nabla \cdot \mathbf{F}$.

   Answer: $2x + 2xz + 2z$

   Difficulty: 1   Section: 1

4. If $\mathbf{F} = (x^2 - y^2)\,\mathbf{i} + 2xyz\,\mathbf{j} + z^2\,\mathbf{k}$, find $\nabla \times \mathbf{F}$.

   Answer: $-2xy\,\mathbf{i} + (2yz + 2y)\,\mathbf{k}$

   Difficulty: 1   Section: 1

5. If $\mathbf{F} = xyz^2\,\mathbf{i} + xe^{yz}\,\mathbf{j} - y\cosh xz\,\mathbf{k}$, find $\nabla \cdot \mathbf{F}$.

   Answer: $yz^2 + xze^{yz} - xy\sinh xz$

   Difficulty: 1   Section: 1

6. If $\mathbf{F} = xyz^2\,\mathbf{i} + xe^{yz}\,\mathbf{j} - y\cosh xz\,\mathbf{k}$, find $\nabla \times \mathbf{F}$.

   Answer: $(\cosh xz - xye^{yz})\,\mathbf{i} + (2xyz - yz\sinh xz)\,\mathbf{j} + (e^{yz} - xz^2)\,\mathbf{k}$

   Difficulty: 1   Section: 1

7. If $\phi = 2x^2yz - z\cos x + e^{yz}$, find $\nabla\phi$.

   Answer: $(4xyz + z\sin x)\,\mathbf{i} + (2x^2z + ze^{yz})\,\mathbf{j} + (2x^2y - \cos x + ye^{yz})\,\mathbf{k}$

   Difficulty: 1   Section: 1

8. If $\phi = 2x^2yz - z\cos x + e^{yz}$, find $\nabla\times(\nabla\phi)$.

   Answer: $\mathbf{0}$ (zero vector)

   Difficulty: 1   Section: 1

9. If $\mathbf{F} = x^2\,\mathbf{i} + y^2\,\mathbf{j} + xyz\,\mathbf{k}$, compute $\nabla \cdot (\nabla \times \mathbf{F})$.

   Answer: 0 (zero scalar)

   Difficulty: 1   Section: 1

10. If $\mathbf{F} = (2x - 5y)\mathbf{i} + (-5x + 3y^2)\mathbf{j} + \mathbf{k}$, find $\nabla \times \mathbf{F}$.

    Answer: $\mathbf{0}$ (zero vector)

    Difficulty: 1   Section: 1

11. If $\mathbf{F} = (6xy^3 + 2z^2)\mathbf{i} + 9x^2y^2\mathbf{j} + (4xz + 1)\mathbf{k}$, find $\nabla \cdot \mathbf{F}$.

    Answer: 0 (zero scalar)

    Difficulty: 1   Section: 1

12. Compute $\int_C yz\,dx + zx\,dy + xy\,dx$ where C is the line segment from $(0,0,1)$ to $(2,3,4)$.

    Answer: 24

    Difficulty: 1   Section: 2

13. Compute $\int_C (x^2 - y)\,dx + (y^2 + x)\,dy$ from $(0,1)$ to $(1,2)$ along the curve $C: y = x^2 + 1$.

    Answer: 2

    Difficulty: 1   Section: 2

14. Evaluate $\int_C \mathbf{F} \cdot d\mathbf{r}$ where $\mathbf{F} = 3x^2\mathbf{i} + (2xz - y)\mathbf{j} + z\mathbf{k}$ and C is the path given by $x = t^2$, $y = t$, $z = 2t^2 - t$, $0 \leq t \leq 1$.

    Answer: $\dfrac{13}{10}$

    Difficulty: 1   Section: 2

15. Evaluate $\int_C \mathbf{F} \cdot d\mathbf{r}$ where $F = 3x^2\mathbf{i} + (2xz - y)\mathbf{j} + z\mathbf{k}$ and C is the straight line from $(0,0,0)$ to $(1,1,1)$.

    Answer: $\dfrac{5}{3}$

    Difficulty: 1   Section: 2

16. Evaluate $\int_C \mathbf{F} \cdot d\mathbf{r}$ where $\mathbf{F} = (y^2 \cos x + z^3)\mathbf{i} + (2y \sin x - 4)\mathbf{j} + (3xz^2 + 2)\mathbf{k}$ and C is the straight line from $(0,0,0)$ to $\left(\dfrac{\pi}{2}, -\dfrac{1}{2}\right)$.

    Answer: $9 + 4\pi$

    Difficulty: 1   Section: 2

17. Evaluate $\int_C \mathbf{F} \cdot d\mathbf{r}$ where $\mathbf{F} = (x^2 - 2xy)\,\mathbf{i} + (x^2y + 3)\,\mathbf{j}$ and C is the path $8y = x^2$ from $(-4, 2)$ to $(4, 2)$.

    Answer: $\dfrac{128}{3}$

    Difficulty: 1   Section: 2

18. Evaluate $\int_C (x^2 + y)\,ds$ where C is the path $x = 3t$, $y = 4t$, $0 \leq t \leq 1$.

    Answer: 25

    Difficulty: 1   Section: 2

19. Evaluate $\int_C (x^2 + y^2)\,ds$ where C is the path $x = e^t \sin t$, $y = e^t \cos t$, $0 \leq y \leq 2$.

    Answer: $\dfrac{\sqrt{2}}{3}(e^6 - 1)$

    Difficulty: 1   Section: 2

20. Is $\mathbf{F} = (2x - 5y)\,\mathbf{i} + (-5x + 3y^2)\,\mathbf{j}$ conservative? If so, find f such that $\mathbf{F} = \nabla f$.

    Answer: Yes, a potential function is $x^2 - 5xy + x^3$

    Difficulty: 1   Section: 3

21. Is $\mathbf{F} = (2x + 5y)\,\mathbf{i} + (2y - 4)\,\mathbf{j}$ conservative? If so, find f such that $\mathbf{F} = \nabla f$.

    Answer: Yes, a potential function is $x^2 + 5x + y^2 - 4y$

    Difficulty: 1   Section: 3

22. Is $\mathbf{F} = (4xy^3 + 17ye^{xy})\,\mathbf{i} + (6x^2y^2 + 17xe^{xy})\,\mathbf{j}$ conservative? If so, find f such that $\mathbf{F} = \nabla f$.

    Answer: Yes, a potential function is $2x^2y^3 + 17e^{xy}$

    Difficulty: 1   Section: 3

23. Is $\mathbf{F} = (3x^2 - 3y^2 - 2x)\,\mathbf{i} + (-6xy + 2y)\,\mathbf{j}$ conservative? If so, find f such that $\mathbf{F} = \nabla f$.

    Answer: Yes, a potential function is $x^3 - 3xy^2 - x^2 + y^2$

    Difficulty: 1   Section: 3

24. Find f so that $\mathbf{F} = \nabla f$ for $\mathbf{F} = (6x^3 + 2z^2)\,\mathbf{i} + (9x^2y^2)\,\mathbf{j} + (4xz + 1)\,\mathbf{k}$.

    Answer: $3x^2y^3 + 2xz^2 + z$

    Difficulty: 1   Section: 3

25. Find f so that $\mathbf{F} = \nabla f$ for $\mathbf{F} = (y^2 \cos x + z^3)\,\mathbf{i} + (2y \sin x - 4)\,\mathbf{j} + (3xz^2 + 2)\,\mathbf{k}$.

Answer: $y^2 \sin x + xz^3 - 4y + 2x$

Difficulty: 1   Section: 3

26. Is $\mathbf{F} = (2xy - y^3 + 2)\,\mathbf{i} + (3xy^2 - x^2 - 4)\,\mathbf{k}$ conservative? If so, find f such that $\mathbf{F} = \nabla f$.

    Answer: No

    Difficulty: 1   Section: 3

27. Is $\mathbf{F} = (3x - 4y)\,\mathbf{i} + (4x - 3y)\,\mathbf{j}$ conservative? If so, find f such that $\mathbf{F} = \nabla f$.

    Answer: No

    Difficulty: 1   Section: 3

28. If $\mathbf{F} = (2x + 5)\,\mathbf{i} + (2y - 4)\,\mathbf{j}$, compute $\int_C \mathbf{F} \cdot d\mathbf{r}$ where C is any piecewise smooth curve from $(0, 1)$ to $(1, 2)$.

    Answer: 5

    Difficulty: 1   Section: 3

29. If $\mathbf{F} = (2x - 5y)\,\mathbf{i} + (-5x + 3y^2)\,\mathbf{j}$, compute $\int_C \mathbf{F} \cdot d\mathbf{r}$ where C is any piecewise smooth curve from $(0, 1)$ to $(1, 2)$.

    Answer: -2

    Difficulty: 1   Section: 3

30. If $\mathbf{F} = (6xy^3 + 2z^2)\,\mathbf{i} + (9x^2y^2)\,\mathbf{j} + (4xz + 1)\,\mathbf{k}$ compute $\int_C \mathbf{F} \cdot d\mathbf{r}$ where C is any piecewise smooth curve from $(0, 1, 0)$ to $(1, -1, 1)$.

    Answer: 0

    Difficulty: 1   Section: 3

31. If $\mathbf{F} = 6xy\,\mathbf{i} + (3x^2 - 3y^2z^2)\,\mathbf{j} - 2y^3z\,\mathbf{k}$ compute $\int_C \mathbf{F} \cdot d\mathbf{r}$ where C is any piecewise smooth curve from $(1, 1, 2)$ to $(0, 1, 3)$.

    Answer: -8

    Difficulty: 1   Section: 3

32. Use Green's Theorem to evaluate $\oint_C (3x - 4y)\,dx + (4x - 2y)\,dy$ where C is the path counterclockwise around the ellipse $x^2 + 4y^2 = 16$ beginning and ending at $(4, 0)$.

    Answer: 64π

Difficulty: 1   Section: 4

33. Use Green's Theorem to evaluate $\oint_C (5x - 2y)\, dx + (2x + 3y)\, dy$ where C is the path counterclockwise around the ellipse $x^2 + 4y^2 = 16$ beginning and ending at $(4, 0)$.

Answer: 32π

Difficulty: 1   Section: 4

34. Use Green's Theorem to evaluate $\oint_C (5x - 2y)\, dx + (-2x + 3y)\, dy$ where C is the path counterclockwise around the ellipse $x^2 + 4y^2 = 16$ beginning and ending at $(4, 0)$.

Answer: 0

Difficulty: 1   Section: 4

35. Use Green's Theorem to evaluate $\oint_C 2xy\, dx + \left(e^x + x^2\right) dy$ where C is the boundary of the triangle with vertices $(0,0)$, $(1,0)$, and $(1,1)$, oriented clockwise.

Answer: 1

Difficulty: 1   Section: 4

36. Use Green's Theorem to evaluate $\oint_C x^2 y\, dx + xy\, dy$ where C is the boundary of the rectangle with vertices $(1,1)$, $(3,1)$, $(1,2)$, and $(3,2)$, oriented clockwise.

Answer: $-\dfrac{17}{3}$

Difficulty: 1   Section: 4

37. Use Green's Theorem to evaluate $\oint_C y\, dx + 2x\, dy$ where C is the boundary of the rectangle with vertices $(0,0)$, $(0,1)$, $(1,0)$, and $(1,1)$, oriented clockwise.

Answer: 1

Difficulty: 1   Section: 4

38. Use Green's Theorem to evaluate $\oint_C \left(2xy - 2x^2 e^y\right) dx + \left(x^2 - x^3 e^y\right) dy$ where C is the circle $x^2 + y^2 = 4$ oriented clockwise.

Answer: 0

Difficulty: 1   Section: 4

39. Use Green's Theorem to evaluate $\oint_C x\left(x^2+y^2\right)^{-1} dx + y\left(x^2+y^2\right)^{-1} dy$ where C is the circle $(x-5)^2 + (y-5)^2 = 4$ oriented clockwise.

Answer: 0

Difficulty: 1   Section: 4

40. Compute $\iint_G x\, dS$ where G is the surface of that portion of the cylinder $x^2 + z^2 = 16$ inside the cylinder $x^2 + y^2 = 16$ in the first octant.

Answer: 32

Difficulty: 2   Section: 5

41. Compute $\iint_G x\, dS$ where G is the surface of that portion of the cylinder $x^2 + z^2 = 4$ inside the cylinder $x^2 + y^2 = 4$ in the first octant.

Answer: 4

Difficulty: 2   Section: 5

42. Compute $\iint_G x\, dS$ where G is the surface of that portion of the cylinder $x^2 + z^2 = 1$ inside the cylinder $x^2 + y^2 = 1$ in the first octant.

Answer: $\dfrac{1}{2}$

Difficulty: 2   Section: 5

43. Compute $\iint_G \mathbf{F}\cdot\mathbf{n}\, dS$ where G is the portion of the surface $3x = 9 - z^2 - y^2$ above the plane $z = 0$, and $F = x\mathbf{i} + 2y\mathbf{j} + 2z\mathbf{k}$.

Answer: $\dfrac{135\pi}{2}$

Difficulty: 2   Section: 5

44. Compute $\iint_G \mathbf{F}\cdot\mathbf{n}\, dS$ where G is the portion of the plane $6z + 3y + 2x = 6$ in the first octant and $F = 6z\mathbf{i} + (2x+y)\mathbf{j} - x\mathbf{k}$.

Answer: 3

Difficulty: 2   Section: 5

45. Use Gauss' Theorem to compute $\iint_G \mathbf{F}\cdot\mathbf{n}\, dS$ where $\mathbf{F} = (x - y^2)\mathbf{i} + (y + 3xz^2)\mathbf{j} + 5z\mathbf{k}$ and

G is the surface consisting of the hemisphere $x^2 + y^2 + z^2 = 4$, $z \geq 0$, and the place surface $z = 0$, $x^2 + y^2 = 4$.

Answer: $\dfrac{112\pi}{3}$

Difficulty: 2   Section: 6

46. Use Gauss' Theorem to compute $\iint_G \mathbf{F} \cdot \mathbf{n}\, dS$ where $\mathbf{F} = (x - 2y)\,\mathbf{i} + (2y - 3z)\,\mathbf{j} + (3z - x)\,\mathbf{k}$ and G is the surface of the cube bounded by $x = 0$, $y = 0$, $z = 0$, and $x = 1$, $y = 1$, $z = 1$.

Answer: 6

Difficulty: 2   Section: 6

47. Use Gauss' Theorem to compute $\iint_G \mathbf{F} \cdot \mathbf{n}\, dS$ where $F = x^2\,\mathbf{i} + 2xy\,\mathbf{j} + 2z\,\mathbf{k}$ and G is the surface of the region in the first octant bounded by $x = 0$, $y = 0$, $z = 0$, and $x + y + z = 1$.

Answer: $\dfrac{1}{2}$

Difficulty: 2   Section: 6

48. Use Gauss' Theorem to compute $\iint_G \mathbf{F} \cdot \mathbf{n}\, dS$ where $\mathbf{F} = (x + e^{-y}\sin z)\,\mathbf{i} + (x^2 + \tan^{-1} z)\,\mathbf{j} + (y \cos x)\,\mathbf{k}$ and G is the surface of the region bounded by the cylinder $z = 4 - x^2$, the plane $y + z = 5$, $z = 0$, and $y = 0$.

Answer: $\dfrac{544}{15}$

Difficulty: 2   Section: 6

49. Use Gauss' Theorem to compute $\iint_G \mathbf{F} \cdot \mathbf{n}\, dS$ where $\mathbf{F} = (xe^{-y}\sin z)\,\mathbf{i} + (x^2 + \tan^{-1} x)\,\mathbf{j} + (y \cos x)\,\mathbf{k}$ and G is the surface of the region above the plane $z = 2$ and below the hemisphere $x^2 + y^2 + z^2 = 8$.

Answer: $\dfrac{8\pi}{3}\left(4\sqrt{2} - 5\right)$

Difficulty: 2   Section: 6

50. Use Gauss' Theorem to compute $\iint_G \mathbf{F} \cdot \mathbf{n}\, dS$ where $\mathbf{F} = (xz - 3\cos y)\,\mathbf{i} + (4x - 3z^2)\,\mathbf{j} + (5xy - \sin x)\,\mathbf{k}$ and G is the surface of the region between $x^2 + y^2 + z^2 = 1$ and $x^2 + y^2 + z^2 = 9$, $z \geq 0$.

Answer: 20π

Difficulty: 2   Section: 6

51. Use Gauss' Theorem to compute $\iint\limits_G \mathbf{F} \cdot \mathbf{n}\, dS$ where

$\mathbf{F} = (4x - yz^2)\,\mathbf{i} + (5y + xz^3)\,\mathbf{j} + (4xy\cos y)\,\mathbf{k}$
and G is the surface of the region in the first octant bounded by the cylinder $x^2 + y^2 = 4$ and the plane $z = 1 + x + 2y$.

Answer: $9\pi + 72$

Difficulty: 2   Section: 6

52. Use Gauss' Theorem to compute $\iint\limits_G \mathbf{F} \cdot \mathbf{n}\, dS$ where

$\mathbf{F} = 3xz\,\mathbf{i} + xyz\,\mathbf{j} + (x^2 + y^2)\,\mathbf{k}$ and G is
the surface of the region in the first octant bounded by $z^2 + x^2 = 9$ and $x + y = 3$.

Answer: $\dfrac{324}{5}$

Difficulty: 2   Section: 6

53. Use Stoke's Theorem to compute $\iint\limits_G (\nabla \times \mathbf{F}) \cdot \mathbf{n}\, dS$ where

$\mathbf{F} = \dfrac{y\,\mathbf{i} - x\,\mathbf{j}}{x^2 + y^2 + z^2}$ and G is the surface
of the hemisphere $x^2 + y^2 + z^2 = a^2$, $z \geq 0$.

Answer: -2π

Difficulty: 2   Section: 7

54. Use Stoke's Theorem to compute $\oint_C \mathbf{F} \cdot \mathbf{T}\, ds$ where C is the curve bounding the portion of the cylinder $x^2 + z^2 = 4$ with $0 \leq y \leq 2$ and $z \geq 0$, and $\mathbf{F} = (x^2 - yz)\,\mathbf{i} + y^2\,\mathbf{j} + (z^2 + xy)\,\mathbf{k}$.

Answer: 8π

Difficulty: 2   Section: 7

55. Use Stoke's Theorem to compute $\oint_C \mathbf{F} \cdot \mathbf{T}\, ds$ where C is the curve bounding the portion of the cylinder $x^2 + z^2 = 1$ with $0 \leq y \leq 2$ and $z \geq 0$, and $\mathbf{F} = (x^2 - yz)\,\mathbf{i} + y^2\,\mathbf{j} + (z^2 + xy)\,\mathbf{k}$.

Answer: 2π

Difficulty: 2   Section: 7

56. Use Stoke's Theorem to compute $\oint_C \mathbf{F} \cdot \mathbf{T}\, ds$ where C is the curve bounding the portion of the cylinder $x^2 + z^2 = 6$ with $0 \leq y \leq 2$ and $z \geq 0$, and $F = (x^2 - yz)\,\mathbf{i} + y^2\,\mathbf{j} + (z^2 + xy)\,\mathbf{k}$.

Answer: 12π

Difficulty: 2   Section: 7

57. Use Stoke's Theorem to compute $\oint_C \mathbf{F} \cdot \mathbf{T}\, ds$ where C is the curve of the intersection of the surfaces $z = 4xy$ and $x^2 + y^2 = 9$, and $F = xz^2\,\mathbf{i} + xz^2\,\mathbf{j} - 4xyz\,\mathbf{k}$.

    Answer: 0

    Difficulty: 2   Section: 7

58. Use Stoke's Theorem to compute $\oint_C \mathbf{F} \cdot \mathbf{T}\, ds$ where C is the curve of the intersection of the surfaces $x^2 + y^2 = 1$ and $x + y + z = 1$, $F = y^2\,\mathbf{i} + x^2\,\mathbf{j} + xy\,\mathbf{k}$.

    Answer: 0

    Difficulty: 2   Section: 7

15 Differential Equations

1. Solve the differential equation $y'' - 3y' - 4y = 0$.

   Answer: $y = Ae^{4x} + Be^{-x}$

   Difficulty: 1   Section: 1

2. Solve the differential equation $y'' + 3y' - 4y = 0$.

   Answer: $y = Ae^{-4} + Be^{x}$

   Difficulty: 1   Section: 1

3. Solve the differential equation $y'' + 2y' - 8y = 0$.

   Answer: $y = Ae^{2x} + Be^{-4x}$

   Difficulty: 1   Section: 1

4. Solve the differential equation $y'' + 2y' + y = 0$.

   Answer: $y = Axe^{-x} + Be^{-x}$

   Difficulty: 1   Section: 1

5. Solve the differential equation $y'' + 4y' + 4y = 0$.

   Answer: $y = Axe^{-2x} + Be^{-2x}$

   Difficulty: 1   Section: 1

6. Solve the differential equation $y'' - 4y' + 5y = 0$.

   Answer: $y = Ae^{2x} \cos x + Be^{2x} \sin x$

   Difficulty: 1   Section: 1

7. Solve the differential equation $y'' + 4y = 0$.

   Answer: $y = A \cos 2x + B \sin 2x$

   Difficulty: 1   Section: 1

8. Solve the differential equation $y'' - 4y = 0$.

   Answer: $y = A \cosh 2x + B \sinh 2x$ or $y = Ae^{2x} + Be^{-2x}$

   Difficulty: 1   Section: 1

9. Solve the differential equation $\left(D^3 - 4D^2 + 6D\right) y = 0$.

   Answer: $y = A + e^{2x}\left(B \sin \sqrt{2}x + C \cos \sqrt{2}x\right)$

   Difficulty: 1   Section: 1

10. Solve the differential equation $(D^2+4)(D-2)(D-1)y = 0$.

    Answer: $y = e^{2x}(A\sin 2x + B\cos 2x) + e^x(C\sin 2x + D\cos 2x)$

    Difficulty: 1   Section: 1

11. Solve the differential equation $y'' - 3y' - 4y = 4x^2 + 6x - 10$.

    Answer: $y = Ae^{4x} + Be^{-x} + 2 - x^2$

    Difficulty: 2   Section: 2

12. Solve the differential equation $y'' + 3y' - 4y = 4x^2 - 6x - 10$.

    Answer: $y = Ae^{-4x} + Be^x + 2 - x^2$

    Difficulty: 2   Section: 2

13. Solve the differential equation $y'' + 2y' - 8y = 4\cos 2x - 12\sin 2x$.

    Answer: $y = Ae^{2x} + Be^{-4x} + \sin 2x$

    Difficulty: 2   Section: 2

14. Solve the differential equation $y'' + 2y' + y = -25\sin 2x$.

    Answer: $y = Axe^{-x} + Be^{-x} + 4\cos 2x + 3\sin 2x$

    Difficulty: 2   Section: 2

15. Solve the differential equation $y'' + 4y' + 4y = 2e^{-2x}$.

    Answer: $y = Axe^{-2x} + Be^{-2x} + x^2 e^{-2x}$

    Difficulty: 2   Section: 2

16. Solve the differential equation $y'' - 4y' + 5y = \sin x - \cos x$.

    Answer: $y = Ae^{2x}\cos x + Be^{2x}\sin x + \dfrac{1}{4}\sin x$

    Difficulty: 2   Section: 2

17. Solve the differential equation $y'' + 4y = \cos 2x - \sin 2x$

    Answer: $y = A\cos 2x + B\sin 2x + \dfrac{1}{4}x(\cos 2x + \sin 2x)$

    Difficulty: 2   Section: 2

18. Solve the differential equation $y'' - 4y = 4e^{2x}$.

    Answer: $y = A\cosh 2x + B\sinh 2x + xe^{2x}$ or $y = Ae^{2x} + Be^{-2x} + xe^{2x}$

    Difficulty: 2   Section: 2

19. Solve the differential equation $y'' - 4y' + 5y = \sin x - \cos x$ if $y(0) = 1$ and $y'(0) = 0$.

Answer: $y = e^{2x}\cos x - \dfrac{9}{4}e^{2x}\sin x + \dfrac{1}{4}\sin x$

Difficulty: 2 Section: 2

20. Solve the differential equation $y'' + 4y = \cos 2x - \sin 2x$ if $y(0) = 0$ and $y'(0) = 1$.

Answer: $y = \dfrac{3}{8}\sin 2x + \dfrac{1}{4}x(\cos 2x + \sin 2x)$

Difficulty: 2 Section: 2

21. A spring with a spring constant k of 96 pounds per foot is loaded with a 12 pound weight and brought to equilibrium. It is then stretched 1 inch and released. Find the motion and the period. Neglect friction.

Answer: $\dfrac{1}{12}\cot 16t$, period $\dfrac{\pi}{8}$

Difficulty: 2 Section: 3

22. A spring with a spring constant k of 64 pounds per foot is loaded with a 16 pound weight and brought to equilibrium. It is then stretched 1 inch and released. Find the motion and the period. Neglect friction.

Answer: $\dfrac{1}{12}\cos 8\sqrt{2}\,t$, period $\dfrac{\sqrt{2}\pi}{8}$

Difficulty: 2 Section: 3

23. A spring with a spring constant k of 18 pounds per foot is loaded with a 12 pound weight and brought to equilibrium. It is then stretched 2 inches and released. Find the motion and the period. Neglect friction.

Answer: $\dfrac{1}{6}\cos 4\sqrt{3}\,t$, period $\dfrac{\sqrt{3}\pi}{6}$

Difficulty: 2 Section: 3

24. A 12 pound weight stretches a spring 2 inches. The weight is raised 2 inches and given an initial velocity of 2 feet per second upward. Find the equation of motion.

Answer: $y = -\dfrac{1}{6}\cos 8\sqrt{3}\,t - \dfrac{\sqrt{3}}{12}\sin 8\sqrt{3}\,t$

Difficulty: 2 Section: 3

25. A 12 pound weight stretches a spring 3 inches. The weight is raised 2 inches and given an initial velocity of 2 feet per second upward. Find the equation of motion.

Answer: $y = -\dfrac{1}{6}\cos 8\sqrt{2}\,t - \dfrac{\sqrt{2}}{8}\sin 8\sqrt{2}\,t$

Difficulty: 2 Section: 3

26. A 12 pound weight stretches a spring 4 inches. The weight is raised 2 inches and given an

initial velocity of 2 feet per second downward. Find the equation of motion.

Answer: $y = -\dfrac{1}{6}\cos 4\sqrt{6}t - \dfrac{\sqrt{6}}{12}\sin 4\sqrt{6}t$

Difficulty: 2   Section: 3

27. A spring with constant k of 10 pounds per foot is loaded with a 32 pound weight and brought to equilibrium. It is then stretches 1 inch and released. Find the equation of motion if the damping force is proportional to twice the velocity.

Answer: $y = e^{-t}\left(\dfrac{1}{12}\cos 3t + \dfrac{1}{36}\sin 3t\right)$

Difficulty: 2   Section: 3

28. A spring with constant k of 10 pounds per foot is loaded with a 32 pound weight and brought to equilibrium. It is then stretches 2 inches and released. Find the equation of motion if the damping force is proportional to twice the velocity.

Answer: $y = e^{-t}\left(\dfrac{1}{6}\cos 3t + \dfrac{1}{18}\sin 3t\right)$

Difficulty: 2   Section: 3